INTERIOR DESIGN OF WORLD BRAND HOTELS

世界顶级酒店室内设计

（荷）柯林·芬尼根 编　常文心 译

辽宁科学技术出版社
沈阳

Hotel Icons, Destinations and Brands
酒店地标、景点和品牌

As a generation, we are fortunate to live in extraordinary times. We live in a global society which is increasingly mobile and international in its outlook. Most of us now have the opportunity to travel anywhere in the world with air travel placing virtually any destination within reach.

With the end of the Cold War and the fall of the Iron Curtain in 1989, a new age of building and restoration commenced, together with opportunities for travel. Cities are redefining their identities, and global markets must now compete with each other to be better places to live, work and visit. Cities are striving to become more beautiful and inviting, and governments and city officials have realised that the key to success can only be achieved through well planned regeneration strategies, with the environment in mind. Take for example the gradual transformation of Germany's Ruhr area. Once associated with its vast industry, the region is now known for its museums, entertainment and various other attractions.

What was previously unthinkable, has in the current urban age become the possible. Cities are transforming themselves into landscapes of pleasure, in which hotels play a central role. Long gone are the days of selecting a hotel for its beds alone, or simply its location. Destination choices involve considering both the place and the experience offered during the stay, be it at a city hotel, a mountain resort or a tropical island. Hotels have therefore become an integral part of the traveller's experience and sometimes itself the reason to visit.

Hotels have helped reshape cityscapes and have altered the international images of cities and countries. They have sometimes defined cities by establishing new icons. We only need to think of Dubai's Burj al Arab hotel, the Raffles Hotel in Singapore or the Waldorf Astoria in New York to recall the essence of the city. Similarly, the Pera Palace Hotel in Istanbul was built as the destination hotel for travellers arriving on the Orient Express, when the city was still known as Constantinople.

Moreover, hotels will always remain influential in branding and forming the image of a city. Rivalry between the various hotel brands has established a greater choice of hotel types and any city typically holds a great number of hotel options. Luxury, business and boutique hotels all cater to the different desires of the contemporary global traveller.

Within city regeneration, hotels can play an important role. Destination hotels act as magnets when opened, creating waves of development in the surrounding area. The conversions of old prisons into luxury hotels or redesigns of abandoned harbour sheds into luxury destinations have become symbolic of the regeneration of neglected city areas. Consider New York's Meat Packing district, or Sydney Harbour's Woolloomooloo Wharf to understand the possibilities for transformation.

Hotels can also play a leading role in rural conservation. Through eco-tourism, vast areas of our planet are being preserved though well managed and enlightened tourist policies. The hotel can be a key factor in a successful conservation effort. An exceptional example of a positive contribution which can be made are the treetop lodges in

Botswana's Okavango Delta.

We at FG Stijl have witnessed the successful regeneration of a city centre site which a building can make, through our renovation of a former 1893 school building in the heart of Amsterdam. The College Hotel, completed in 2005 and awarded the Prix Villégiature Paris for 'Best Hotel Interior Design in Europe' (2005), has revitalised a corner of the city and also provides educational facilities for the hotel industry.

FG Stijl's design philosophy and motto is "designed by designers, built for people". We try to create projects worthy of a journey, and most importantly with a sense of place. We take our inspiration first and foremost from the location, views and local culture and our interiors and architecture reflect this approach. We concentrate on logical spatial planning and through this create unique experiences and places to be. What really distinguishes projects from one another is the layering of design factors. This could be the level of environmental qualities, local skills or historical importance. The experience for the guest is defined by the combination of these factors, the atmosphere created through the detailing of the work, all of which make the visitor's journey worth the effort.

A new and soon to be familiar iconic view is the Golden

Horn Bay waterfront at Vladivostok. Located on the eastern seaboard of Russia and also the end station of one of the world's most famous train journeys, Vladivostok now takes six days to reach on the high speed Trans-Siberian Express across the vast expanses of Russia. Its unique geographical location places Vladivostok only a couple of hours by plane from China or Japan, positioning it as the most European city in Asia. Founded by the Tzars of Russia, Vladivostok retains much of the nineteenth century architecture and was for many years closed to the outside world. It lies within a region where the Amur Tigers, the Far Eastern Leopard and the Himalayan Black Bear roam freely.

The Golden Bay, local culture and architecture provided a rich source of inspiration for FG Stijl in the design of the Hyatt Regency Golden Horn hotel. Until recently there was not a definitive world-wide image representing the city; however, it is now fixed in the process of major redevelopment and regeneration, thanks to it being chosen to host the 24th summit of the Asia Pacific Economic Cooperation forum (APEC). Massive infrastructure projects include two dramatic bridges which connect the peninsulas of the Golden Horn Bay to Vladivostok. This hotel project is integral to the regeneration process and occupies an iconic location at the heart of the harbour on the site of the original pier, affording convenient moorings for super-yachts, and spectacular views of the harbour mouth and evening sunsets.

Recent city developments have gravitated towards the establishment of urban and world icons. Each world icon influences the manner through which global societies associate with cities. Brands want to be associated with icons, so the association of city, icon and branding seems to be a natural effect.

Hotel brands today are recognised globally. It is our view that design can make the brand not only stronger but also relevant to the site. Hotels must be at the centre of any city's contemporary branding and image strategy, and those within the quality of known brands can meaningfully express the individual location. In other words, global brands need to emphasise the local by way of tying down how global brands meet local places. The hotel experience can help to differentiate one destination from another and for the discerning traveller, provide not only expected levels of comfort, but also an experience which is truly memorable.

FG Stijl
Colin Finnegan | Gerard Glintmeijer

我们这一代幸运地生活在非凡的时代，全球化社会越来越具有移动和国际化特点。我们大多数人都有机会乘飞机到世界各地的景点旅行。1989年，冷战结束，东欧剧变，开启了建造和修复的新时代，也为旅行提供了机会。城市正重新树立自己的形象，在全球市场中相互竞争，成为更适合居住、工作和旅行的地点。城市正努力变得更加美丽和吸引人，而政府和市政官员们也认识到了成功的关键在于计划周详的重建策略，当然，环境也是重要因素。例如，德国鲁尔地区的逐步改造。该地区已经从工业基地成功地转型为以博物馆、娱乐和其他景点而著称的旅游胜地。

从前难以想象的一切，在现在的城市中都变成了可能。城市正逐步改造成为愉悦的景观，而酒店在其中起到了重要的作用。酒店早就已经超越了住宿和便利的基本功能。旅行者在选择酒店的过程中会考虑酒店自身和它所提供的服务体验，是选择城市酒店、山庄度假村还是热带岛屿。因此，酒店已经成为了旅行者体验的一部分，甚至是他们选择旅行的原因。

酒店帮助重塑了城市景观并改变了城市和国家的国际形象。它们有时会通过建立新地标来重新定义城市形象。迪拜的帆船酒店、新加坡的莱佛士大酒店以及纽约的华尔道夫酒店都巩固了城市的地位。同样

的，早在城市还被称作君士坦丁堡的时候，伊斯坦布尔的皇宫酒店就已经成为了到达东方快车的旅行者的首选酒店。

此外，酒店一直对城市品牌影响力和形象的树立有着至关重要的影响。各种酒店品牌的竞争促生了许多酒店类型，几乎每座城市都有许多酒店选择。奢华酒店、商务酒店和精品酒店满足了现代全球旅行者的不同需求。

在城市重建过程中，酒店能够起到重要的作用。酒店就像磁石一样，会引发周边区域的开发浪潮。将监牢进行改造或是重新设计废弃的港口仓库作为奢华酒店是典型的城市重建行为。纽约的肉类加工区、悉尼港的伍卢穆卢码头都经历了这种改造。

酒店在乡村田园保护中同样扮演了重要角色。生态旅游让我们星球广阔的区域通过旅游政策得到了保护。酒店是保护过程中的决定性因素。博茨瓦纳奥卡万多三角洲的树屋酒店就是一个具有积极作用的典范。

在对阿姆斯特丹一座建于1893年的学校教学楼的翻修过程中，FG风格工作室见证了一次成功的城市中心重建。学院酒店完工于2005年，获得了2005年巴黎度假奖"欧洲最佳酒店室内设计奖"。酒店复兴了城市的一角，将教学设施改造为酒店业所用。

FG风格工作室的设计哲学和座右铭是"设计师来设计，建筑以人为本"。我们设法打造值得一游的项目，使其具有独特的地方感。我们首先从项目位置、视野和当地文化中获得灵感，并通过建筑和室内设计反映出来。我们专注于逻辑空间规划，并以此来创造独一无二的空间体验。真正让项目脱颖而出的是设计元素的层次。这包括环境质量等级、地方技巧或是历史重要性。宾客的体验由这些元素综合决定，整体氛围则由细部设计所营造。这一切都让旅客的旅行物有所值。

海参崴的金角湾打造了一个全新的地标式酒店。海参崴坐落在俄罗斯的东海岸，是世界著名火车旅行线路的终点站。从海参崴乘坐高速跨西伯利亚快车需要六天来横跨俄罗斯广阔的领土。独特的地理位置让海参崴乘飞机到中国或日本仅需几小时，是亚洲最重要的欧洲城市。由俄国沙皇所建造的海参崴保持了19世纪的建筑风貌，过去的多年一直与世隔绝。阿穆尔虎、远东豹和喜马拉雅黑熊都在海参崴自由漫步。

在凯悦金角湾酒店的设计中，黄金湾、当地文化和建筑为FG风格工作室提供了丰富的设计灵感。多年以来，海参崴并没有一个固定的城市形象；作为第24届亚太经贸合作组织峰会的主办国，海参崴正经历着大规模重建和再开发，其中包括两座连接金角湾和海参崴的大桥。酒店项目对重建进程至关重要，占据着港口码头的中心位置，方便游艇停泊，同时拥有港口和落日的壮丽美景。

新近的城市开发都趋向于城市和世界地标性建筑的建立。每个世界地标都影响着全球化社会与城市的联系方式。各大品牌都希望与地标建筑相联系，让城市、地标建筑和品牌形象看起来像是一种自然效应。酒店品牌得到了全球的认可。我们认为设计不仅能够让品牌更强大，还会影响其所在地。酒店必须是城市现代品牌化和形象策略的中心，同时，知名品牌的品质还会表现独特地点的意义。也就是说，全球品牌需要通过与本地的联系来凸显地区的特色。酒店体验能够区分不同的景点；对挑剔的旅行者来说，它不仅提供了预期的舒适，还是一次真正令人难忘的体验。

FG风格工作室
柯林·芬尼根｜杰勒德·格林特梅贾尔

ACCOR 雅高

012 SOFITEL GUANGZHOU SUNRICH
广州圣丰索菲特大酒店

026 SOFITEL MACAU AT PONTE 16
澳门十六浦索菲特酒店

036 SOFITEL PHNOM PENH PHOKEETHRA
RESORT & SPA
金边索菲特佛基拉度假村

STARWOOD 喜达屋

048 THE ST. REGIS BANGKOK
曼谷瑞吉酒店

060 THE ST. REGIS ROME
罗马瑞吉酒店

070 THE ST. REGIS FLORENCE
佛罗伦萨瑞吉酒店

080 SHERATON BANGALORE AT BRIGADE
GATEWAY
班加罗尔Brigade Gateway喜来登酒店

090 SHERATON HUIZHOU BEACH RESORT
惠州金海湾喜来登度假酒店

THE PENINSULA 半岛

100 THE PENINSULA HONG KONG
香港半岛酒店

108 THE PENINSULA TOKYO
东京半岛酒店

122 THE PENINSULA BANGKOK
曼谷半岛酒店

INTERCONTINENTAL 洲际

134　INTERCONTINENTAL PUXI

　　　上海浦西洲际酒店

144　INTERCONTINENTAL REGENCY BAHRAIN

　　　巴林洲际酒店

FOUR SEASONS 四季

152　FOUR SEASONS LOS ANGELES AT BEVERLY HILLS

　　　洛杉矶比佛利山四季酒店

162　FOUR SEASONS HOTEL DENVER

　　　丹佛四季酒店

MARRIOTT 万豪

172　RITZ-CARLTON HONG KONG

　　　香港丽思卡尔顿酒店

188　RITZ-CARLTON DUBAI IFC

　　　迪拜丽思卡尔顿酒店

196　JW MARRIOTT HOTEL BEIJING

　　　北京万豪酒店

210　JW MARRIOTT MARQUIS MIAMI

　　　迈阿密万豪伯爵酒店

HILTON 希尔顿

220 THE SKIRVIN HILTON OKLAHOMA CITY

希尔顿斯科文俄克拉何马城酒店

230 HILTON GUANGZHOU TIANHE

广州天河新天希尔顿酒店

236 HILTON CHENNAI

金奈希尔顿酒店

244 WALDORF–ASTORIA SHANGHAI ON THE BUND

上海外滩华尔道夫酒店

HYATT 凯悦

254 HYATT REGENCY DUSSELDORF

杜塞尔多夫凯悦酒店

266 GRAND HYATT MACAU

澳门君悦酒店

276 HYATT REGENCY JING JIN CITY RESORT AND SPA

京津新城凯悦酒店

286 PARK HYATT SEOUL

首尔柏悦酒店

294 PARK HYATT BEIJING

北京柏悦酒店

304 GRAND HYATT GUANGZHOU

广州富力君悦大酒店

312 HYATT REGENCY HONG KONG, TSIM SHA TSUI

香港尖沙咀凯悦酒店

322 PARK HYATT SHANGHAI

上海柏悦酒店

332 HYATT REGENCY HANGZHOU

杭州凯悦酒店

340 HYATT REGENCY HONG KONG, SHA TIN

香港沙田凯悦酒店

MANDARIN ORIENTAL 文华东方

350   MANDARIN ORIENTAL, TOKYO

东京文华东方酒店

364   MANDARIN ORIENTAL, SINGAPORE

新加坡文华东方酒店

372   MANDARIN ORIENTAL, BOSTON

波士顿文华东方酒店

382   THE LANDMARK MANDARIN ORIENTAL,

HONG KONG

香港置地文华东方酒店

SHANGRI-LA 香格里拉

392   SHANGRI-LA HOTEL, GUANGZHOU

广州香格里拉大酒店

402   SHANGRI-LA HOTEL, WENZHOU

温州香格里拉大酒店

412   SHANGRI-LA HOTEL, XI'AN

西安香格里拉大酒店

424   SHANGRI-LA HOTEL, SUZHOU

苏州香格里拉大酒店

434   SHANGRI-LA HOTEL, GUILIN

桂林香格里拉大酒店

442   SHANGRI-LA HOTEL TOKYO

东京香格里拉大酒店

450   SHANGRI-LA'S FAR EASTERN PLAZA HOTEL,

TAINAN

香格里拉台南远东国际大饭店

460   KERRY HOTEL PUDONG, SHANGHAI

上海浦东嘉里大酒店

476   INDEX

索引

ACCOR HOTELS GROUP
A Vogue-classic French Complex

雅高酒店集团———时尚古典的法式情怀

As the largest hotel group in Europe, Accor Hotels Group processes 4,000 hotels and nearly 45,000 rooms all around the world, spreading 90 countries in five continents. The proud "Accor spirit" is the supporting core of the hotels' development. This artistic hotel spirit combines historical traditions and modern innovations in the hotels' architectural and interior design, creating fancy sparkles between French tone and local cultures.

Nowadays, Accor Hotels are developing rapidly in Europe and Asia. The delicate hotels have become the top options of travellers for their excellent architectural and interior designs, which lies in their serious considerations on the hotel design. Accor Hotels have invited numerous well-known architects and interior designers for their hotel designs, including the 2008 Pritzker winner – French architect Jean Nouvel, Moroccan interior designer Karim Chakor and French interior designer Didier Gomez, even the Fashion designer Kenzo Takada and Christian Lacroix.

Accor Hotels not only emphasise industrialisation and standardisation, but also require high-level design. Different brands of Accor Hotels have different design definitions according to their brand definitions. For instance, the luxurious Sofitel hotels prefer the combination of history and fashion. Many Sofitel hotels are restored from historical architectures with hundreds of years of history. In 2009, Sofitel created Sofitel Legend. Located in a historic scenic spot, each hotel retains the palace-like façade. With the matching lighting effect, the hotel looks like a jewellery in the historical remains. The interior design has a unified style. For example, the public areas and guestrooms need to keep the fashionable or classic styles. The design of different functional areas emphasises its distinctive characteristic. For instance, the lobby of a Sofitel hotel will highlight the luxurious and exquisite atmosphere of the hotel. Normally, the design is based on grand Chinese or European classic style, with some themed or innovative decorations to emphasise the visual shock. Besides, a lobby always pays attention to its lighting design to create a sense of high-quality, comfort and chic.

Furthermore, in the interior designs of the Accor hotels, we could perceive the hotels' high pursuit of style and art through the ubiquitous French tones. The guestrooms are delicate and elegant, with some decorations of art works and flowers and some fashionable furniture. Sofitel SO Boutique hotels is a brand of French styles. Each SO hotel will invite some well-known designers even fashion designers to create a unique fashion atmosphere. Sofitel SO Mauritius invited fashion master Kenzo to design the guestrooms, and therefore fashion is seen everywhere from decorations to fabrics. Of course, the Sotitel hotel will never forget its love for French vogue. Sofitel SO Bangkok invited a Thai architect and five local designers and integrated the chic elements of French fashion master Christian Lacroix to express the respect to him and French culture.

Through the display of three Sofitel hotels of Accor Group, we wish to provide readers with an opportunity to understand the brand hotels' architectural and interior designs. Let's explore how the leading hotel brand creates French atmosphere in the pursuit of vogue and classics.

雅高酒店集团是欧洲第一大酒店集团，在全世界拥有 4,000 家酒店和近 45,000 个房间。酒店足迹遍布五大洲 90 个国家。支持酒店发展的是雅高引以为豪的"雅高精神"，这堪称综合艺术的酒店精神在酒店的建筑与室内设计之中，将历史的传统与现代的创新融合，让法式情调与本土文化碰撞出新奇的火花。

雅高酒店目前在欧洲与亚洲发展迅速，座座精致的酒店凭借优秀的建筑与室内设计成为各地的食宿首选。这与他们对酒店建筑与室内设计的重视是分不开的。在雅高酒店的建造师以及室内设计名单上可谓星光熠熠，这其中包括 2008 年普利兹克奖得主———法国建筑师 Jean Nouvel，以及摩洛哥室内设计师 Karim Chakor 与法国室内设计师 Didier Gomez，甚至还会看到时尚设计师 Kenzo Takada 和 Christian Lacroix。

追求产业化、标准化的雅高酒店在酒店的设计方面也有着高的要求。纵观各雅高酒店，根据各品牌酒店的定位，都有着基本的设计定义。例如具有世界奢华级水准的索菲特酒店，有着对历史与时尚的特殊偏爱。很多索菲特酒店翻新于历史建筑，有着上百年的历史。2009 年，索菲特创立索菲特传奇酒店，每一座酒店都选址在历史名胜地，配合着城市的古典景观，索菲特保留了宫殿般的外立面，配合以渲染的灯光效果，让其更像遗迹中的珍宝。室内的整体风格也有统一的要求，包括公共区域、客房等功能区应保持或时尚现代，或古典的风格。具体在设计各个功能区时针对不同区域的特点也有专门的侧重点，例如索菲特酒店的大堂都会突出酒店豪华精致的气氛，通常以宏伟的中式或欧式古典风格为基础，再加以主题的描绘或装饰创新，突出视觉上的震撼。另外，大堂一般都注重灯光的独特设计，以此来赋予大堂的高品质感、舒适感及时尚感。

另外，在雅高酒店的室内设计中，弥漫着的法式情调让人可以看出酒店对时尚与艺术的超高追求。酒店的客房精致而优雅，适当的装点艺术品与鲜花，并布置时尚的家具。索菲特 SO 酒店系列，这是一系列法式风格精品酒店。每一座 SO 酒店都会邀请著名的设计师甚至时尚设计师来打造专属的时尚氛围。例如毛里求斯的索菲特 SO 酒店邀请时尚大师 Kenzo 参与布置酒店客房，从酒店的装饰品到布艺都能看到时尚的影子。而弘扬法式情调的索菲特当然也不会忘记表达对法式时尚的钟爱之情，曼谷索菲特 SO 酒店邀请一位泰国建筑师和 5 位本土设计师，在对酒店的设计中融入了法国时尚大师 Christian Lacroix 的时尚符号，以表达对这位大师及法国文化与艺术的敬仰。

本章通过对雅高旗下 3 家索菲特酒店的展示，希望能提供给读者一个近距离了解本品牌酒店建筑与室内设计的机会。看看酒店业内追求时尚与古典的巨子如何在世界营造法式情怀的。

SOFITEL GUANGZHOU SUNRICH

广州圣丰索菲特大酒店

Ideally located in the heart of Tianhe, Guangzhou's most dynamic financial and business district, the Sofitel Guangzhou Sunrich is less than one kilometre from the emblematic skyscraper CITIC Plaza and a few minutes from the East Railway Station, a veritable hub that connects Guangzhou to Dongguan, Shenzhen and Hong Kong.

"Sofitel Guangzhou Sunrich is going to breathe a little of the unique French culture into Guangzhou. The goal of this address is to become the meeting place for the city's cultural events and especially through all forms of Art," declared Robert Gaymer-Jones, CEO Sofitel Worldwide.

The hotel offers 493 rooms and suites, whose décor elegantly blends contemporary Asian design and Parisian chic. All rooms overhang the characteristic skyline of the Tianhe district. Everyone is pampered right into their dreams in the exclusive softness of the MyBed™ bedding signed Sofitel.

Among the five bars and restaurants, the fine dining restaurant "Robata Grill & Bar", which combines a classy steakhouse with modern Japanese Izakaya, features an authentic Robatayaki, a sushi bar with an extensive elegant walk-through wine cellar. "Le Chinois" highlights famous Cantonese gastronomy, "2 on 988" a gourmet all-day dining restaurant with five open kitchens, raw stations, wood fire oven, authentic French rotisserie, traditional Chinese kitchen, Western kitchen, pastries and bakeries. The "8 Faubourg" reproduces around an elegant bar the décor typical of a Parisian apartment. The "Mar-Tea-Ni" lounge bar is the inescapable place to savour fine French pastries accompanied by rare teas suggested by a Tea Sommelier. Like this, Sofitel creates a link between French culture and Chinese culture, a value dear to the brand.

Completion date: June, 2011
Location: Guangzhou, China
Designer: CCD

Photographer: Hotel Team
Area: 73,816m²

完成时间：2011 年 6 月
项目地点：中国，广州
设计师：CCD

摄影师：酒店团队
面积：73,816 平方米

1. The luxurious lobby in the hotel shows French artful and cultural atmosphere.
2. With the custom pendant lights and carpet, Mar Tea Ni located in the lobby offers Art Décor ambience.

1. 酒店奢华的大堂展示了法式艺术和文化氛围
2. 定制的吊灯和地毯彰显了大堂酒吧独特的装饰艺术氛围

广州圣丰索菲特大酒店位于天河区——广州最活跃的金融商业区,距离中信广场不足1千米,并且紧邻连接广州和东莞、深圳、香港的火车东站。

索菲特集团全球总裁罗伯特·盖米尔–琼斯说:"广州圣丰索菲特大酒店将为广州带来独特的法国文化。酒店的目标是成为城市文化活动、特别是各种艺术活动的会议场所。"

酒店拥有493间客房和套房,它们的装饰优雅地结合现代亚洲设计和巴黎风尚。所有房间都能享有天河区的独特景色,而MyBed品牌床品将以其独一无二的柔软带领客人进入美梦。

在酒店的五家餐厅和酒吧中,"六福宫"日本餐厅,结合经典牛排餐厅和现代日式居酒屋,以正宗的炉边烧——一个配有高雅的可通过式酒窖的寿司吧——为特色。"南粤宫"以著名的广东美食为特色,"2 on 988"全日制餐厅拥有五间开放式厨房、生肉台、木炭烤箱、正宗的法式烤肉、传统中国小厨、西方小厨、面食和面包店。"巴黎8号"的设计采用了巴黎公寓的经典装饰。"Mar Tea Ni"酒廊是品味法式精致面点和稀有茶饮的好去处。索菲特酒店在法国文化和中国文化之间建立了联系,这对该品牌来说异常珍贵。

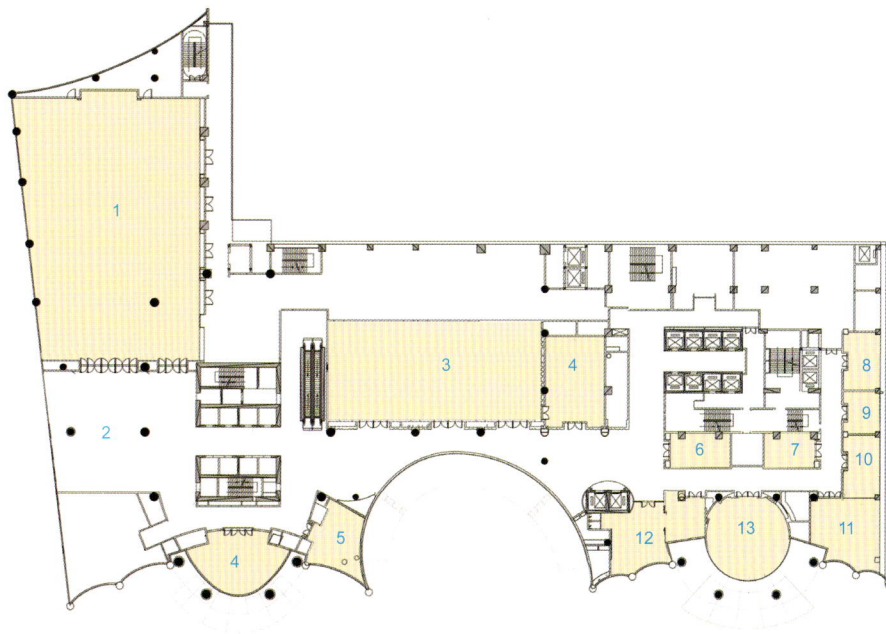

1. Grand ballroom   1. 圣丰大宴会厅
2. Pre-function area   2. 活动区域
3. Salle Victor Hugo   3. 雨果厅
4. VIP   4. 贵宾厅
5. Salon Renoir   5. 雷诺瓦厅
6. Salon Bizet   6. 比才厅
7. Salon Le Notre   7. 勒诺特厅
8. Salon Ravel   8. 拉威尔厅
9. Salon Bejart   9. 贝嘉厅
10. Salon Doisneau   10. 杜瓦诺厅
11. Salon Monet   11. 莫奈厅
12. Business centre   12. 商务中心
13. Salon Flaubert   13. 福楼拜厅

1. The "Mar-Tea-Ni" lounge bar is the inescapable place to savour fine romantic atmosphere.
2. In Club Millesime, the interior is inspired by Chinese brush drawing.

1. 大堂酒吧拥有精致浪漫的氛围
2. 年度葡萄酒俱乐部的室内设计从中国画中获得了灵感

1. The Cantonese cuisine restaurant – Le Chinois proposes a journey through authentic home style.
2. One of five open kitchens in "2 on 988"
3. "Robata Grill & Bar" also features a sushi bar with an extensive elegant walk-through wine cellar.

1. 全日制广式餐厅具有正宗的私房菜风格
2. 2 on 988餐厅五间开放式厨房中的两间
3. 六福岛餐厅的寿司吧配有巨大而优雅的酒架

2

3

1. Wooden lattice, round table, old Chinese chairs, carpet with flower motif, all of these in the private dining room of Le Chinois are full of oriental styles.

2. The meeting room features plush sofas and carpet.

3. The spacious pre-function area in the meeting room

1. 木窗格、圆桌、中式椅、花式地毯，全日制餐厅包房充满了东方气息

2. 会客室以舒适的沙发和地毯为特色

3. 会客室宽敞的准备区

1. In ballroom, the colour palette of Chinese red and royal yellow creates a plush space for any meeting or banquet.

2. The meeting room is equipped with state-of-the-art lighting equipment.

3. The conference room for small meeting combines both traditional furnishing and modern amenities.

1. 宴会厅的中国红和皇室黄为不同类型的会议和宴会打造了奢华的空间

2. 会议室配有先进的灯光设备

3. 小型会议室融合了传统装饰和现代设施

1. The décor in guestroom elegantly blends contemporary Asian design and Parisian chic.
2. The bed upholstered with silk is matched with the carpet with French words.
3. Red in Chinese culture means a lot, which is used throughout the whole hotel as a motif. The red Eiffel Tower in the painting is telling a story named "When French art meets Chinese culture".
4. The separated bathroom in the suite presents extreme luxury.

1. 客房的装饰结合了现代亚洲设计与巴黎风尚
2. 床上铺着丝绸床品，与带有法文的地毯相互搭配
3. 红色在中国文化里意味深长，贯穿了整个酒店；画中的红色埃菲尔铁塔意味着"法国艺术与中国文化的邂逅"
4. 套房的独立浴室极致奢华

SOFITEL MACAU AT PONTE 16

澳门十六浦索菲特酒店

Positioned on Macau's picturesque waterfront, in the centre of the charming historic quarter, with a walking distance to the 25 UNESCO enlisted world heritage sites, perfect location for shopping, food and culture hunting, Sofitel Macau is ideal for international travellers savouring the glamorous mix of chic and modern styles.

The 408 rooms — with 19 VIP Mansions — of refined elegance, exquisite restaurant, bar, recreational facilities, Club lounge, meeting and banquet venues with WiFi access and MJ gallery ensure that Sofitel has created an exquisite world for true lovers of "art de vivre".

Mansion at Sofitel outlines uniqueness and exclusivity. The magnificent Mansion at Sofitel offers the ultimate in style, elegance and personal comfort. With four different design themes, the 19 units of mansion provide the finest amenities and facilities for convenience and peace of mind when travelling.

The Mansion at Sofitel is furnished with state-of-the-art technology, rain forest shower and Jacuzzi. The generous space, stylish décor, tremendous views of Macau peninsula and Pearl River Delta at dawn and dusk, and above all, personalised butler services, all made up the most memorable luxury experience!

The 120 to 262 sqm mansions with one-to-three bedroom set ups, from one floor to duplex, featuring Eternal Glamour, Black Galaxy, Blanc Romance and Avant Garde.

Eternal Glamour translates glamour into chic and contemporary style. Living room features leather lounges. Crystal décor conveys elegance in the right tone. Ensuite master bedroom with walk-in wardrobe, spacious bathroom in luxurious marble design with Jacuzzi and rainforest shower make up a comfort zone.

Black Galaxy sees the classical moulding of imperial design and décor, noble styling. Ensuite master bedroom features oversized luxurious bathroom with Jacuzzi, rain sky shower, massage lounge and sauna room. Entertainment room is featured.

Blanc Romance as named is a design of an all-in-white feminine romance! The white motif colour and glass furniture with polyhedral crystals reflecting cleanness and unity, bringing a sense of tranquility and prestige.

Avant Garde, the tasteful designed duplex reads the concept of "homey" yet "trendy", a cup of tea for young professionals! The contemporary designed tea & coffee table at the living room, the arty carpet, Phillipe Starck's dining chairs and decorative items are delicate and eye-catching. The classical touch of the Italian designed freestanding bathtub looks just right for the character of the spacious bathroom!

Completion date: 2008
Location: Macau, China
Designer: The Jerde Partnership, Richards Basmajian

Limited (The Mansion at Sofitel)
Photographer: Sofitel Macau at Ponte 16
Area: 23,000m²

完成时间：2008 年
项目地点：中国，澳门
设计师：捷得建筑师事务所；理查德·倍斯马吉安

公司（濠庭十六浦）
摄影师：澳门十六浦索菲特酒店
面积：23,000 平方米

澳门索菲特酒店坐落在澳门历史中心城区风景如画的水畔，距离25项世界文化遗产仅有几步之遥，周边环境适合购物、就餐和文化狩猎。酒店完美地结合了时尚和现代风格，是国际旅客休息入住的绝佳场所。

408间优雅的客房（其中包括19套贵宾公寓）、精致的餐厅、酒吧、娱乐设施、酒廊、配有无线网络的会议及宴会场所和米高积逊珍品廊让索菲特酒店成为了"生活的艺术"的爱好者的天堂。

濠庭十六浦以独特感和专属感为特色。索菲特的豪宅在风格、典雅和个人舒适感上做到了极致。19个公寓单元氛围四个不同的主题，将为客人在旅途中提供最精致的设施和心灵的平静。

索菲特公寓配有最先进的技术设施、热带雨林花洒和按摩浴缸。宽敞的空间、时尚的装饰、澳门半岛的宏大视野和珠江三角洲的晨昏美景以及个性化的管家服务，共同组成了难忘的奢华体验。

120到262平方米的公寓配有一到三间卧室，既有单层公寓也有双层公寓，分别采用了四个主题："Eternal Glamour"、"Black Galaxy"、"Blanc Romance"和"Avant Garde"。

"Eternal Glamour"主题将魅力转化为时尚而现代的风格。客厅以皮革沙发为特色。水晶装饰传递出优雅的情调。套房主卧室配有步入式衣柜，而宽敞的大理石浴室内设有按摩浴缸和雨林花洒。

"Black Galaxy"主题展示了皇室设计和贵族风格的经典。套房主卧室以特大的奢华浴室为特色，浴室内设有按摩浴缸、雨淋花洒、按摩休息室和桑拿房。卡拉OK房是它的特色。

"Blanc Romance"主题采用了女性化的全白色浪漫设计。白色图案色彩和玻璃家具反映出清洁感和整体感，为空间带来了宁静感和豪华感。

"Avant Garde"主题将双层套房打造成居家而时尚的感觉，适合年轻的专业人士的品位。客厅里拥有现代设计感的咖啡桌、艺术地毯、菲利普·史塔克设计的餐椅和装饰品精致而引人注目。意大利设计的独立浴缸正适合宽敞的浴室。

1. Entering the lobby, the exquisite interior is highlighted by crystal ball pendants.
2. The splendid Rendezvous lounge in the lobby features noble furnishing.

1. 酒店大堂的精致装饰在水晶球吊灯的点缀下更加突出
2. 大堂酒廊的装饰高贵典雅

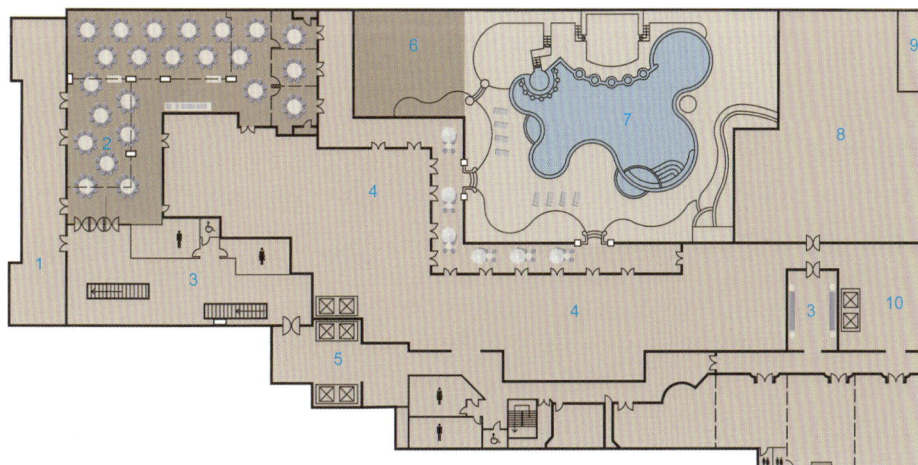

1. South Terrace
2. Baccara multi-function room
3. Foyer
4. Mistral Restaurant
5. Guest lifts
6. Le Terrace
7. Swimming Pool
8. Fitness Centre
9. Changing room with steam and sauna room
10. Mansion lifts

1. 百家乐多功能厅的露天平台
2. 百家乐多功能厅
3. 门廊
4. 海风餐厅
5. 客用电梯
6. 多功能厅外的露天平台
7. 游泳池
8. 健身中心
9. 更衣室：桑拿及蒸汽浴室
10. 濠庭十六浦电梯

1. The lotus pond in the atrium savours the sunshine through the glass ceiling. The fairy flower is the design motif throughout the hotel.
2. The popular Macau meeting venue features state-of-the-art equipment and facilities, sunset river views, creative theme dinners and extensive car parking.

1. 中庭的莲花池从玻璃天花板汲取阳光；莲花图案贯穿了酒店设计
2. 这个在澳门深受欢迎的宴会场所以先进的设施、落日河景、创意主题晚宴和充足的停车位为特色

1. The living room at Black Galaxy features European style furnishing and chandelier.
2. Avant Garde, the tasteful designed duplex reads the concept of "homey" yet "trendy".
3. Blanc Romance as named is a design of an all-in-white feminine romance.

1. 黑色银河套房的客厅以欧式风格装饰和吊灯为特色
2. 先锋设计跃层套房兼具居家感和时尚感
3. 白色浪漫采用了女性化的全白色浪漫设计

1

1. Black Galaxy sees the classical molding of imperial design and décor, noble styling.

2. In Blanc Romance, the white motif colour and glass furniture with polyhedral crystals reflect cleanness and unity, bringing a sense of tranquility and prestige.

3. Eternal Glamour translates glamour into chic and contemporary style. Crystal décor conveys elegance in the right tone.

4. Spacious bathroom in Eternal Glamour is in luxurious marble design with Jacuzzi and rainforest shower making up a comfort zone.

1. 黑色银河套房体现了皇室设计和装饰，风格尊贵

2. 白色浪漫套房的白色图案色彩和玻璃家具显得简洁而完整，带来了宁静感和尊崇感

3. 永恒魅力套房呈现出时髦而现代的风格；水晶装饰传递出优雅的感觉

4. 永恒魅力套房内宽敞的浴室采用了奢华的大理石设计，配有极可意按摩浴缸和雨林淋浴设施

SOFITEL PHNOM PENH PHOKEETHRA RESORT & SPA

金边索菲特佛基拉度假村

Sofitel Phnom Penh Phokeethra is set riverside amongst pristine landscaped gardens in the old quarter of the city. The colonial style hotel is conveniently located within close proximity to key attractions, embassies and Phnom Penh's central business district. The hotel's luxurious 201 rooms and suites boast views of either the Mekong or Bassac rivers. An array of restaurants and bars can be found within the hotel as well as an upscale bar, two swimming pools, a sports club and state-of-the-art conference facilities, the finest in Cambodia.

The hotel is designed and inspired by Choochart Polakit with a French colonial style. All rooms and suites are spacious and well appointed with wooden floors. The hotel presents the French contemporary flair mixed with Cambodian architecture and provides all facilities to maximise the guests' satisfaction.

The Sofitel Phnom Penh Phokeethra is a delightful haven blending Asian and European elements while enjoying a business lunch or sophisticated retreat.

The hotel's striking lobby with high coffered ceilings, wrought iron chandeliers and polished marble floors, is reminiscent of a 1920s' French colonial estate. The intricate wood latticework, rattan furniture and ceiling fans evoke the style of Southeast Asia.

Decked out in a natural palette, with dashes of bright colourful accents, rooms and suites lavish you with luxurious and high-tech facilities. The eight restaurants and bars offer a range of delectable cuisine and creative menus that promise to delight your senses.

Sofitel Phnom Penh Phokeethra offers 201 rooms and suites, including 121 superior rooms, 45 luxury rooms, 23 junior suites, 11 prestige suites and the opera suite. All with an intimate elegance of hardwood and a warm palette of colours. All rooms feature a working enclave with the latest technology. Each room is ultimate

Completion date: March, 2011
Location: Phnom Penh, Cambodia
Designer: Choochart Polakit (Thailand)

Photographer: Ukit Hanamornset
Area: 52,000m²

完成时间：2011年3月
项目地点：柬埔寨，金边
设计师：库查尔特·波拉基特（泰国）

摄影师：乌吉特·哈那摩恩赛特
面积：52,000平方米

in decadence with French luxurious bathroom amenities and spacious interiors, while the palatial opera and prestige suites include separate living and lounge areas for relaxing or entertaining.

A dedicated building hosts the Phokeethra Sports Club, fully devoted to the well-being of body and mind. The complex features a fully-equipped fitness centre including an outdoor swimming pool, a multi-activities room, a kids club, two squash courts, and four floodlit tennis courts. A veritable invitation to socialising and friendly interaction while enjoying the calmness of the hotel's large botanical gardens.

Seven state-of-the-art meeting rooms including the Phokeethra Grand Ballroom provide an inspiring environment for a wide range of functions. Luxuriously furnished with high ceilings and ornate crystal chandeliers, the 1,800-sqm Phokeethra Grand Ballroom and the 1,200-sqm pre-function area are suitable for large-scale conferences, exhibitions and the perfect venue for weddings.

1. The hotel's striking lobby with high coffered ceilings, wrought iron chandeliers and polished marble floors, is reminiscent of a 1920s' French colonial estate.
2. All conference rooms are equipped with the latest audio-visual technology and mobile communication systems.

1. 酒店宏伟的大堂装饰着方格天花板、熟铁吊灯和抛光大理石地面，令人回忆起20世纪20年代的法式殖民地建筑
2. 所有会议室都配有最新的视听技术和移动通信系统

1. Entrance
2. Parking
3. Tennis
4. Pool
5. Kids club
6. Sports bar
7. Fitness
8. Squash
9. Sports club
10. Fu Lu Zu – Chinese Fusion Restaurant
11. Le bar – Lobby bar
12. Lobby
13. Chocolat
14. Hachi – Japanese Robatayaki Restaurant
15. Hotel 201 rooms & suites, 5 meeting rooms, Sospa
16. La coupole – All day dining
17. Do Forni – Italian restaurant
18. Aqua bar – Swimming pool bar
19. Ballroom
20. 70 car underground parking

1. 入口
2. 停车场
3. 网球场
4. 游泳池
5. 儿童俱乐部
6. 运动酒吧
7. 健身中心
8. 壁球室
9. 运动俱乐部
10. 中餐厅
11. 大堂酒吧
12. 大堂
13. 巧克力店
14. 日式炉边烧餐厅
15. 201间客房和套房，五间会议室和水疗中心
16. 全日制餐厅
17. 意大利餐厅
18. 游泳池
19. 宴会厅
20. 70个车位的地下停车场

金边索菲特佛基拉度假村坐落在河畔，环绕在老城区的原始景观花园之中。这座殖民地风格的酒店紧邻主要的城市景点、使馆和金边的中央商务区。酒店的201间奢华客房和套房享有湄公河或巴塞河的美景。酒店内设有一系列餐厅和酒吧、一间顶级酒吧、两个游泳池、一家运动俱乐部和先进的会议设施，是柬埔寨最好的酒店。

酒店设计拥有法国殖民地风格。所有客房和套房都配备了木地板。酒店的法式现代风情和柬埔寨建筑相互结合，提供了能够最大化宾客舒适感的全部设施。

金边索菲特酒店是结合了亚洲和欧洲元素的天堂，为宾客提供美味的商务午餐或精致的度假服务。

酒店宏伟的大堂装饰着方格天花板、熟铁吊灯和抛光大理石地面，令人回忆起20世纪20年代的法式殖民地建筑。复杂的木格子、藤条家具和吊扇则显示出东南亚的风情。

客房和套房在自然调色中点缀了明亮的彩色，为你带来了奢华的高科技设施。8间餐厅和酒吧提供了一系列美味的美食和创意菜单，提升了感官的愉悦。

客房：金边索菲特酒店共有201间客房和套房，其中包括121间高级客房、45间豪华客房、23间普通套房、11间豪华套房和一间歌剧套房。所有房间都配有优雅的硬木和温暖的色调，同时配备着最先进的科技设备。每间客房都拥有奢华的法式卫浴设施和宽敞的室内空间，而富丽堂皇的歌剧套房和豪华套房中还设有独立的起居和休息空间，方便休闲和娱乐。

佛基拉运动俱乐部设在一座独立建筑之中，致力于身心健康的修复。设施齐全的健身中心内设有室外游泳池、多功能活动室、儿童俱乐部、两间壁球室和四个照明网球场。俱乐部吸引着人们来此进行社交和友好互动，同时也享受着酒店巨大的植物园的宁静。

7间先进的会议室中包括功能齐全的佛基拉大宴会厅。装饰着豪华的高天花板和华丽的水晶吊灯，1,800平方米的大宴会厅和1,200平方米的准备区适合举办大型会议、展览和婚礼。

1. Phokeethra Grand Ballroom provides an inspiring environment for a wide range of functions. It's furnished luxuriously with high ceilings and ornate crystal chandeliers.
2. La Coupole with stunning wood panelling and high ceilings offers the experience of marvelous show kitchen and the appreciation to perfect simplicity.

1. 佛基拉大宴会厅适合举办各种各样活动；挑高天花板和玛瑙水晶吊灯尽显奢华
2. 库波勒餐厅精妙的木镶板和挑高天花板衬托了展示厨房，整体装饰完美而简单

1. Overlooking Phnom Penh and the mighty Mekong, the executive lounge – Club Millesime is the perfect place for mixing pleasure and business on the rooftop floor with a suspended garden.
2. The entrance to Fu Lu Zu – Chinese restaurant is designed with strong Chinese cultural elements.
3. In Italian restaurant – Do Forni, a wonderful wine cellar is surrounded by natural light, wood ceilings and water ponds.
4. Hachi – Japanese robatayaki restaurant is specialised with the private tatami room.

1. 行政酒廊——年度葡萄酒俱乐部俯瞰着金边和湄公河，在屋顶打造了休闲和商务的乐园，同时配有一个空中花园
2. 福禄寿中餐厅的入口采用浓重的中式文化元素设计
3. 多弗尼意大利餐厅日光充足，采用了木天花板和水池装饰，是个美妙的酒窖
4. 八号日式炉边烧餐厅设有独特的榻榻米包房

1. The palatial opera suite includes separate living and lounge areas, and provides French luxurious experience.
2. The living room in Prestige Suite is designed in rich Colonial style.
3. The bedroom in Prestige Suite features Sofitel Mybed.
4. In Chocolat, a bake shop, a refined bookshelf is equipped as the dining background, adding more cultural atmosphere.

1. 富丽堂皇的歌剧套房内设有独立的起居和休息区，提供了法式奢华体验
2. 声望套房的客厅采用了丰富的殖民地风格装饰
3. 声望套房的卧室配有索菲特专属大床
4. 巧克力烘焙坊内精致的书架作为就餐背景增添了更多的文化氛围

STARWOOD HOTELS AND RESORTS
A Kaleidoscope in Hotel Design Industry

喜达屋酒店集团——酒店设计业的万花筒

starwood
Hotels and
Resorts

As one of the largest three hotel companies in the USA, Starwood Hotels and Resorts owns nearly 1,000 hotels in 100 countries. Besides, Starwood Hotels and Resorts is a hotel company which possesses the largest number of brand names. Its 9 brands can meet any guests' needs all over the world.

The brand names include the following: St. Regis has a long history and emphasise personal service and luxury. The Luxury Collection is a group of unique hotels and resorts offering unique, authentic and enriching experiences indigenous to each destination that capture the sense of both luxury and place. All the hotels have a history of at least 300 years. W is where iconic design and cutting-edge lifestyle set the stage for exclusive and extraordinary experiences. Each hotel and retreat is uniquely inspired by its destination, fashionable and luxurious. Aloft offers a sassy, refreshing, ultra effortless alternative for guests both with its open spaces and infinite opportunities. Element aims to create a balanced atmosphere through fluid multi-functional spaces, possessing a flexible and environmentally friendly design based on nature. Westin pursues luxurious infrastructure and highlights its business atmosphere, trying to provide a brand new hotel experience. Le Méridien combines European traditions and modern cultures, creating an elegant atmosphere with deep cultural accents and humanistic content. Sheraton is the largest brand of Starwood, occupying 40% of Starwood's hotels and Four Points by Sheraton is its sub brand. The various subsidiaries can satisfy the unique requirements from all kinds of guests. Unlike the unified design concept and spirit of the other hotel groups, each brand of Starwood is a distinctive flower. Their different architectural and interior designs enrich and foster the splendid kaleidoscope of Starwood together.

This chapter selects five hotels from two qualified brands – St. Regis and Sheraton to provide a general view of their designs.

Each of the three St. Regis hotels has its own features. St. Regis hotels and resorts spread all over the world, including London, New York, Singapore, and Bali. They typically have individual design characteristics to capture the distinctive personality of each location. The standard of opulence and sophistication established by the original St. Regis is honoured in every address, but has evolved to include five distinct design interpretations: Metropolitan Manor, Glass House, Hemispheres, Journey's End and Paradise Found. In each, the essence of the brand and its rich traditions is brought to life through signature features such as grand staircases, glittering chandeliers, handsome libraries, vast wine vaults, iconic murals and bronze façades.

This chapter displays two Sheraton hotels in the end. The design target of a Sheraton hotel is to promote communication and provide a home-like place. The lobby is intentionally designed as the most appropriate place for communication. The space is designed modern and striking, with some classic charms. Meanwhile, the bedroom is always designed in a warm tone, providing a sense of hospitality. The details are sophisticated yet intimate, creating a sense of comfort. Sheraton provides a modern interpretation for the timeless classics.

喜达屋酒店集团是美国三大酒店集团之一，在全球的 100 个国家拥有近 1,000 家酒店。此外，喜达屋也是全球拥有最多品牌酒店的酒店管理集团，其拥有 9 个酒店品牌，包揽了全球不同客人所需要的任何一种酒店类型。

这9个品牌包括历史久远，强调私人服务的奢华品牌——瑞吉；每家酒店都有300年以上历史，堪称旅行目的地门户，使旅客尽情领略原汁原味的当地文化和无限魅力的豪华精选；每一间酒店都有自己的风格，都被设计、音乐和时尚新闻领域的尖端人物赋予了新的生命，兼具标志性时尚感与现代奢华感的 W 酒店；前卫时尚、清新有趣，注重个性化特色，为客人营造开放式空间并开启无限可能的雅乐轩；旨在通过流畅的多功能空间营造平衡和谐的氛围，拥有自然为本的灵活、环保型设计空间的源宿；追求硬件奢华，突出商务氛围，以为客人打造焕然一新的酒店体验为己任的威斯汀；将欧洲的传统与当代文化融合起来，营造出富有浓郁文化气息、深邃人文内涵典雅氛围的艾美；以及占喜达屋酒店总数达 40% 的高端奢华品牌喜来登 Sheraton 以及附属品牌福朋喜来登 Four Points by Sheraton。全方位，面对不同受众的品牌酒店设置，能满足各类客人的独特体验要求。不像其他大酒店集团统一的设计理念与精神，喜达屋的每一个品牌似一朵独特之花，它们各自的建筑与室内设计风韵，共同丰富，并滋养着喜达屋这只绚烂的万花筒。

本章特别选取最有资历的两大品牌酒店——瑞吉以及喜来登的五间酒店。分别看看这两大品牌酒店的设计风采。

首先是三家最具特色的瑞吉酒店。瑞吉酒店与度假村遍布全球的多个精彩地点。伦敦、纽约、新加坡、巴厘岛——每家酒店都能让您领略到当地的迷人风情和独特风韵。此外，每家酒店还一如既往地承袭了首家瑞吉酒店所确立的豪华与精致标准，并在不断发展中又纳入五种与众不同的设计风格：都市庄园、玻璃屋、精致半球、旅程终点和天堂乐园。每种设计风格均以独特手法完美再现了瑞吉品牌的丰富精髓与悠久传统，例如宽大的楼梯、熠熠闪光的吊灯、雅致的图书馆、琳琅满目的酒窖、标志性壁画和青铜正面装饰。

本章最后展示了两家喜来登酒店。喜来登的设计目标为能够让酒店成为促进交流，并提供如家般舒适的场所。在大堂，空间被刻意设计为最适合人们交流的地方。空间被布置得现代而瞩目，但是仍保留露出经典风韵。同时，卧房通常为暖色调，给人款待之感，设计细节讲究而精致但不缺少亲切的舒适感。可以说，喜来登为永恒的经典提供了一种现代感的诠释。

THE ST. REGIS BANGKOK

曼谷瑞吉酒店

Inspired by the vertical standing stones created by Eastern and Western cultures throughout time, considered a spiritual expression of the human connection to – and reverence towards – the natural world, The St. Regis Bangkok has been designed as a contemporary abstract monolith. Mysterious and awe-inspiring, the seemingly effortless simplicity of the building's silhouette conceals the sophisticated and extensive planning achieved to distil the design to its purest expression.

The St. Regis Bangkok occupies levels 12-24 of a 47-storey mixed-use development in the heart of Bangkok, together with 53 residential units that make up The Residences at St. Regis Bangkok. Exquisite design, commanding views, meticulous attention to detail and uncompromising St. Regis services and amenities ensure an exceptional guest experience.

Appearing austere from afar, upon closer inspection the façades reveal a highly textured marriage of stainless steel, black granite, and glass. White aluminium banding at each floor illustrates the connection between the urban scale of the building and the human scale of its components. Meticulous detailing of design elements including the windows, handrails and canopies brings a gracious touch to The St. Regis Bangkok, creating an emotional connection for the guests from the moment of arrival until their departure.

Entering The St. Regis Bangkok through impressive carved timber and nickel-framed glass doors bearing the St. Regis emblem, guests will be greeted or assisted by waiting staff. To the left and right, The St. Regis Bangkok's signature restaurants are concealed by a Thai screen during the day and revealed dramatically in the evening, when the restaurant is open. On both sides of the entry, guests will find the social lounge, where they can enjoy coffee, tea or cocktails. Guests can wait, work or relax with privacy in the lounge, outfitted with a mix of comfortable sofas and lounge chairs in cream, charcoal and dark plum; seating pods with wing-back chairs; a side console; magazines and newspapers; and decorative reading lamps.

Inspired by Thai basket weaving and made with carved timber and backlit Thai silk, the rear wall of the lobby features deconstructed bronze Thai pots in bronze, illuminated from within for casting shadows on the white Thai silk behind. Additional accessories and artwork are a mix of contemporary and classic Thai.

Completion date: 2011
Location: Bangkok, Thailand
Designer: Brennan Beer Gorman Architects

Photographer: Ralf Tooten
Area: 95,000m²

完成时间：2011 年
项目地点：泰国，曼谷
设计师：BBG 建筑事务所

摄影师：拉尔夫·杜顿
面积：95,000 平方米

Glamorous with high ceilings, rich drapes, elegant sheers and loose rugs on hand-tufted carpets, The Drawing Room or the lobby lounge drops three steps down and offers extraordinary views of the Bangkok sports club. The main bar directly ahead on arrival is designed to be the stage for the St. Regis Bloody Mary and Champagne Sabering Rituals. The solid timber bar boasts an enormous, beautifully framed mirror, animating the space and reflecting the guests into a living mural. Oversized, comfortable sofas sit on rugs flanked by a selection of Asian and contemporary furniture and lights. Exquisitely styled and lavishly appointed, the accommodations redefine luxury in Bangkok. Timeless elegance in design is the hallmark of the 227 guest rooms including 51 suites. Floor-to-ceiling windows offer unobstructed views of the city skyline and of the greenery of the local parks and golf courses. The luxury of space in the rooms ranges from 45 square metres to 250 square metres for the signature suites.

1. Outdoor swimming pool is decked out with strong Thai cultural elements.
2. Guests can wait, work or relax with privacy in the lounge, outfitted with a mix of comfortable sofas and lounge chairs in cream, charcoal and dark plum.
3. The solid timber bar boasts an enormous, beautifully framed mirror, animating the space and reflecting the guests into a living mural. Oversized, comfortable sofas sit on rugs flanked by a selection of Asian and contemporary furniture and lights.

1. 室外游泳池装饰着浓厚的泰式文化元素
2. 休息室为客人提供了等候、工作和休息的私密空间，配有舒适的沙发和长椅，以奶油色、木炭色和深梅红色为主要色调
3. 结实的木制吧台拥有一面巨大的镜子，活跃了整个空间，倒映出客人的活动，形成了一面鲜活的壁画。地毯上放置着超大的舒服沙发，并配有一系列的中式或现代造型的家具与灯饰。

东西方文化中的立石是人类与自然世界之间联系的精神表现，同时也体现了人类对大自然的敬意。曼谷瑞吉酒店从立石中获得了灵感，被设计成一座现代抽象的独立石块建筑。建筑轮廓的朴素感神秘而令人敬畏，将设计过程中所花费的大量努力掩藏了起来。

曼谷瑞吉酒店占据着曼谷市中心一座47层建筑的12-24层，与53套住宅共同组成了曼谷瑞吉公寓大楼。精致的设计、居高临下的视野、一丝不苟的细节和毫不马虎的瑞吉酒店服务和设施保证了宾客们的非凡体验。

从远处看去，建筑似乎十分朴素，毫无装饰。但是走近来看，它的外立面上混合了不锈钢、黑色花岗岩和玻璃所形成的纹理。各个楼层之间的白铝带显示了建筑的城市特征与其所蕴含的人性尺度之间的联系。设计元素一丝不苟的细节（包括窗户、栏杆和天棚）为曼谷瑞吉酒店带来了雅致的风格，为宾客营造了由始至终的情感联系。

华丽的木刻、镍包边玻璃门上装饰着瑞吉酒店的象征标志。从玻璃门进入，宾客将从侍者那里得到热情的招呼和服务。曼谷瑞吉酒店的特色餐厅白天隐藏在泰式屏风之后，晚上则大放异彩。入口两侧均设有社交休息室，客人们可以在那里享用咖啡、茶或鸡尾酒。休息室为客人提供了等候、工作和休息的私密空间，配有舒适的沙发和长椅，以奶油色、木炭色和深梅红色为主要色调。同时，休息室里还提供围座椅、侧面控制台、报纸杂志和装饰性台灯。

受到了泰式平纹编织的启发，大堂的后墙采用了木刻和背光式泰国丝绸进行装饰，呈现出解构的泰国铜水壶图案，背光照明在后面的白色丝绸上投下了阴影。附加的装饰和艺术品混合了现代和古典泰式风格。

配有挑高型天花板、丰富的帷帘、优雅的薄纱和松软的地毯，会客厅和大堂吧拥有曼谷运动俱乐部的非凡景象。正对入口的吧台供应着瑞吉酒店特色的鸡尾酒和香槟。结实的木制吧台拥有一面巨大的镜子，活跃了整个空间，倒映出客人的活动，形成了一面鲜活的壁画。特大号的沙发摆放在地毯上，两侧摆放着亚洲和现代家具和灯具。

精巧的风格和充足的设施让酒店重新定义了曼谷的奢华。永恒的优雅感是227间客房（包括51间套房）的设计特色。落地窗提供了没有阻碍的城市美景和本地公园及高尔夫球场的绿色空间。奢华的客房空间规模从45平方米到250平方米（特色套房），大小不一。

1. By day, Viu's atmosphere reflects the elegance of a stately home, transforming in the evening to a seductive dining venue.
2. The bar offers an elegant and refined ambiance with high ceilings, polished wood floors and a subtle palette of dark, polished wood, charcoal grey and coffee tones, all bathed in soft light.
3. Through the artistic use of sliding panels and ambient lighting, the visually compelling venue creates a changing character and atmosphere in which guests enjoy inviting breakfast, lunch and dinner menus.

1. 白天，维欧餐厅展现出豪宅的典雅；夜晚，它就变成魅惑的就餐空间
2. 在柔和的光线中，挑高天花板、抛光木地板和深木色、炭灰色与咖啡色的结合为酒吧提供了优雅而精致的氛围
3. 拉板和气氛灯光的艺术运用为餐厅打造了变换的特色和氛围，邀请客人来享用早、午、晚餐

1. The 515-square-metre Astor Ballroom is the ideal location for glittering weddings and sophisticated events.
2. The St. Regis Bangkok offers four smaller meeting rooms, Rajadamri I-IV, named for the hotel's legendary location on Rajadamri Road. Two spaces, rooms I and II measure 102 and 105 square metres respectively and face the Royal Bangkok Sports Club.

1. 515平方米的奥斯特宴会厅是举办婚礼和高级宴会的理想场所
2. 曼谷瑞吉酒店的小型会议室以酒店的所在位置——拉加丹瑞路命名，名为拉加丹瑞1号到4号；1号、2号会议室分别为102平方米和105平方米，朝向曼谷皇家俱乐部

1. Pre-function
2. Astor ballroom
3. Meeting room

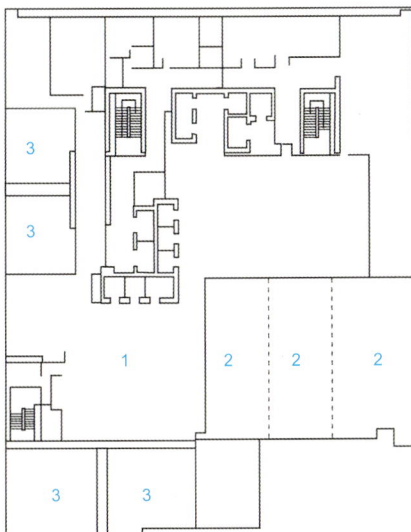

1. 准备区
2. 阿斯特宴会厅
3. 会议室

1. Each ballroom offers a contemporary, elegant space and may include a pre-function space of 282 square metres. The smaller boardroom style meeting rooms III and IV measure 63 and 58 square metres and look out on the vibrant city centre.

2. Furnishings are arranged for intimate conversations as well as just sitting back on a plush banquette or sofa and taking in the stunning views from floor-to-ceiling windows overlooking the Royal Bangkok Sports Club and the dazzling Bangkok Skyline.

3. Viu offers three dining rooms, each with different settings and characters to reflect the time of day. Brightly coloured table mats are changed according to the time of day to match the red and cream coloured uphosltered chairs.

1. 宴会厅营造了现代、优雅的空间，配有282平方米的准备区；3号和4号小型会议室面积分别为63平方米和58平方米，远眺充满活力的城市中心

2. 室内装饰既适合私密的会话又适合静静地休息；从落地窗能够俯瞰曼谷皇室运动俱乐部和曼谷耀眼的天际线

3. 维欧餐厅有三个区域，分别采用不同的布置和特色来反映一天之中不同的时间；色调明亮的桌垫会随着时间的变换而改变，与红色和奶油色的软椅相互搭配

1. With a corner location and panoramic views of the sweeping cityscape and verdant parkland through floor-to-ceiling windows, the distinguished Caroline Astor suites offer a unique, residential style retreat for relaxing or entertaining. Guests will slumber peacefully in the signature St. Regis King bed fitted with lavish 300 thread-count linens, a down comforter, plump pillows and a soft throw.

2. With an exclusive residential style that exudes Thai traditions and contemporary design, each spacious 47-to 55-square-metre Grand Deluxe Room offers an urban to escape for discerning business or leisure travellers.

3. The refined basin in bathroom of guestroom

4. 5. The elegant bathrooms offer extremely comfortable experience.

1. 尊贵的卡洛琳·爱斯特套房拥有转角的有利位置，落地窗展示了城市全景和郁郁葱葱的公园景观，是一个具有独特住宅风格的住所；瑞吉特制大床上铺有奢华的300针织床单、羽绒被、饱满的枕头和柔软的盖毯，客人可以在上面进入安静的睡眠

2. 传统泰式住宅风格和现代设计的运用让47~55平方米的大豪华房为独具慧眼的商务和休闲旅客提供了城市中的净土

3. 客房内精致的洗手池

4、5. 典雅的浴室提供极致的舒服体验

THE ST. REGIS ROME

罗马瑞吉酒店

The project, born in partnership with Polittico di Roma, a famous, contemporary art gallery located in Rome, wants to show the aesthetic and cultural changes which characterised the St. Regis Rome. Thus, visual art works created by international young contemporary artists embellish some of the suites and public areas of the hotel in particular the lobby, Le Grand Bar and Vivendo Restaurant, all inspired by the most significant piazzas and palaces in Rome.

Each of the 138 guestrooms and 23 suites feature a unique identity and are exquisitely appointed in a combination of Empire, Regency and Louis XV styles.

The traditional home of monarchs, heads of state and celebrities, the Royal Suite embodies pure opulence. An entry hall leads to a beautiful living room with a grand piano. The suite also has a master bedroom with king bed, guest bedroom, two dressing closets and a marble bathroom with Jacuzzi. The private dining room is furnished with a marble table that seats 12 and is served by a private kitchen and wine cellar. Louis XVI and Piedmontese furnishings complement the wine red, gold and green colour scheme and a genuine Aubusson carpet is underfoot.

The Bottega Veneta Suite, designed by Tomas Maier and unveiled in October 2007, is a unique and expansive environment created for those who appreciate quiet luxury, unparallelled service and a refined sensibility. This one-of-a-kind suite features a foyer, three bedrooms (two with separate sitting areas), three baths, and a living room with a fireplace. The palette is serene and sophisticated, featuring light colours that create a cool and airy effect.

The Designer Suite is on the ground floor of the hotel and features an entrance hall, living room and a bedroom with bath — the bath features a Jacuzzi — and dressing areas. Coffered ceilings, richly detailed wood and marble floors, sumptuous yet decidedly contemporary architecture, furnishings and fabrics characterise this suite. Among the original artwork from private collections are pieces by important contemporary artists such as Luca Pignatelli, Ubaldo Bartolini and Paolo Fiorentino.

Inspired by the St. Regis passion for fashion, art and literature, the Ambassador Couture Suite has been entirely designed and projected by HBA London, in the spirit of the movie Roman Holidays. Inspired by the city's renowned fashion and design studios, the suite features signature furnishings and accessories which celebrate the best of Italian style such as: a private library

Completion date: 2007
Location: Rome, Italy
Designer: Michael Stelea – HDC Interior Architecture
+ Design, Architect Tomas Maier

Photographer: Starwood Hotels & Resorts
Worldwide, Inc.
Area: 6,800m²

完成时间：2007 年
项目地点：意大利，罗马
设计师：迈克尔·斯特利——HDC 室内设计公司；
托马斯·迈尔

摄影师：喜达屋酒店集团
面积：6,800 平方米

with an anthology of Italian cinema classics; historical photograph of many celebrity guests and a photograph of red smoke chosen for its allusion to Valentino's collection. Situated on the first floor of the St. Regis Rome the 150 sqm suite comprises a lobby, three bedrooms (two with separate sitting areas), three bathrooms, and a living room with a working fireplace and three large windows overlooking the city.

"Vivendo" makes its mark on dining in the capital, just as the hotel did on hospitality and accommodation. The restaurant provides a sophisticated environment with a contemporary atmosphere. The furniture recalls the 30's and 40's style, offering an informal but refined environment, well balanced with the St. Regis Rome, where it is located. Seventy places, an intimate room for 14 guests, the "Champagnerie", lighting designed to help the diners relax, but at the same time to emphasise the works of culinary art offering products typical of the area combined with intense fragrances and aromas.

1. Murano glass chandeliers, antique rugs, vintage furniture, are all joined in the lobby.
2. The reception area features splendid frescoes and carved marble counter.

1. 大堂里装饰着慕拉诺玻璃吊灯、古董地毯、复古家具
2. 前台接待区以璀璨的壁画和大理石台面为特色

项目与罗马著名的现代艺术画廊"罗马祭坛"进行了合作，希望呈现罗马瑞吉酒店的美学和文化价值。因此，一些套房和酒店的公共区域（特别是酒店大堂、酒吧和维万多餐厅）都采用了年轻的国际现代艺术家的视觉艺术作品进行装饰，这些艺术品从罗马著名的广场和宫殿中获得了灵感。

138间客房和23间套房都各有其独特的特色，巧妙地结合了帝国时期、摄政时期和路易十五时期的风格。

作为君主、国家元首和名流的惯例住所，皇家套房尽显奢华。门厅引导着宾客进入放置着大钢琴的华丽客厅。套房还拥有配有特大双人床的主卧、客卧、两间更衣室和一个配有极可意浴缸的大理石浴室。私人餐厅的大理石桌台配有12个座位，由私人厨房和酒窖提供服务。路易十六时期风格和皮埃蒙特风格的家具呈现出酒红、金色和绿色的搭配，地面上铺着正宗的奥布松地毯。

宝缇嘉·温妮达套房由托马斯·迈尔设计，于2007年10月正式开放，为热爱宁静奢华、至尊服务和精致感性的人们提供了独特的宽敞环境。这个独一无二的套房由门厅、三间卧室（其中两间配有独立的起居区）、三间浴室和一间配有壁炉的客厅组成。房间色调宁静而精致，以浅色打造出清爽、轻盈的效果。

设计师套房位于酒店的一楼，由门厅、客厅、配有浴室（以极可意按摩浴缸为特色）的卧室和更衣区组成。方格天花板、精致丰富的木地板和大理石地面、奢华而明晰的现代建筑风格、家具和织物奠定了套房的特色。众多私人收藏的原创艺术品包括卢卡·皮革那特里、乌尔多·巴托里尼和保罗·费洛伦迪诺等著名现代艺术家的作品。

大使时装套房传承了瑞吉酒店对时尚、艺术和文学的热爱之情，完全由HBA伦敦分公司设计，采用了电影《罗马假日》的风格。套房从罗曼著名的时装设计室中获得了灵感，以定制家具和配饰来诠释意大利风格的精髓，例如：私人图书室内收藏着意大利经典电影选集，名流宾客的历史照片以及暗示着华伦天奴品牌的红色烟雾照片。套房位于酒店的二楼，总面积150平方米，配有三间卧室（其中两间配有独立的起居区）、三间浴室、一间配有壁炉的客厅和三扇俯瞰城市景色的大窗户。

维万多餐厅专注于餐饮，正如酒店专注于服务和住宿。餐厅在现代气氛中提供了精致优雅的环境。餐厅的家具摆设采用20世纪30、40年代复古风格，打造出放松而精致的环境，与罗马瑞吉酒店的整体风格十分相称。餐厅可容纳70人就餐，还有一个可容纳14人的私人包房——香槟房。灯光设计帮助就餐者们放松，同时也凸显了香气四溢的当地美食。

3

1. An intimate room for 14 guests, the "Champagnerie" in Vivendo, lighting designed to help the diners relax.
2. Le Grand Bar is the ideal place to relax and take in the atmosphere of a historic landmark in Rome.
3. The furniture in Vivendo restaurant recalls the 30's and 40's style, offering an informal but refined environment.

1. 香槟酒廊是一个可容纳14人的私密空间，灯光营造出休闲的就餐氛围
2. 大酒吧是休闲的好去处，拥有罗马特有的历史气息
3. 维万多餐厅的家具采用了20世纪三四十年代的风格，打造了休闲而精致的空间

1. Interior features stones from Diocletian's Baths. Guests enjoy intimate dinners among the world's finest wines in di...Vino Private Wine Cellar.
2. The Salone Ritz, the first ballroom inaugurated in Rome has returned to its antique splendour, thanks to the arduous restoration of its splendid frescoes by the famous painter, Mario Spinetti.

1. 私人酒窖的设计从皇帝浴场中获得了灵感，客人们可以在里面品尝世界顶级美酒
2. 里兹大厅是罗马第一个重返古时显赫的宴会厅，拥有著名画家马里奥·史宾那提绘制的华丽壁画

1. The Royal Suite embodies pure opulence. An entry hall leads to a beautiful living room with grand piano. Louis XVI and Piedmontese furnishings complement the wine red, gold and green colour scheme and a genuine Aubusson carpet is underfoot.

2. Junior Suites feature soft velvet armchairs, sofas and curtains in warm tones of red, gold and beige, offering exceptional comfort and elegance.

3. Each Ambassador Suite is personalised by a different colour scheme and refined décor and amenities throughout; the effect is warm and elegant. All are available with one king, queen or twin bed. Precious damask fabrics, Murano glass chandeliers and lamps, original frescos and stucco give a distinct flair to each suite.

1. 皇室套房富丽堂皇；门厅通往配有大钢琴的华丽客厅；路易十六和皮埃蒙特装饰采用了酒红色、金色和绿色，地上铺着欧比松地毯

2. 普通套房配有柔软的天鹅绒扶手椅、沙发和窗帘，以红色、金色和米黄色为主色调，显示出非凡的舒适和优雅

3. 每间大使套房都拥有独特的色彩搭配和装潢，显得温馨而优雅；套房配有一张特大的双人床或两张单人床；珍贵的锦缎织物、慕拉诺玻璃吊灯和灯具、原创设计的壁画和灰泥装饰让每间套房都拥有独特的韵味

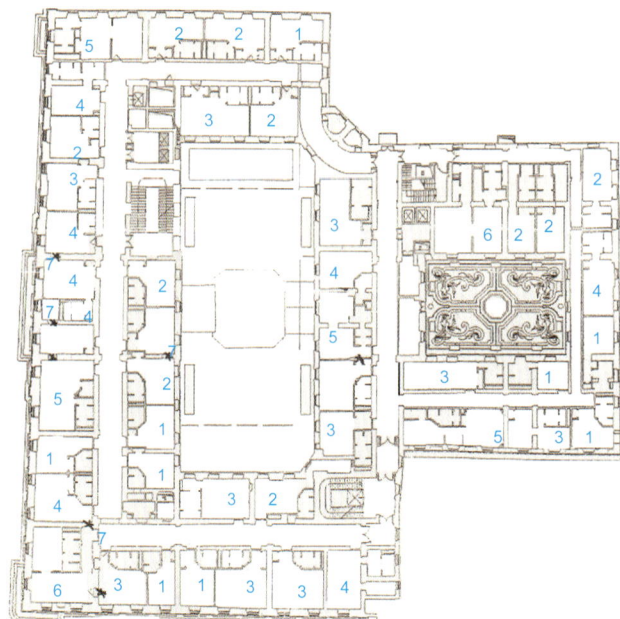

1. Superior
2. De Luxe
3. Imperial
4. Twin
5. Junior Suite
6. Ambassador Suite
7. Connections

1. 高级客房
2. 豪华客房
3. 皇室客房
4. 双子客房
5. 普通套房
6. 外交套房
7. 连接点

THE ST. REGIS FLORENCE

佛罗伦萨瑞吉酒店

The St. Regis Florence features 81 luxuriously appointed guest rooms and 19 suites, including a spectacular designed suite by Bottega Veneta. Reflecting Florence's rich artistic heritage, all the hotel's guest rooms and suites feature individually hand-carved gold leaf plaques including one of three different colour palettes – Medici, Florentine and Renaissance, with each design concept offering a unique selection of custom-designed furniture, paintings, frescos and crystal chandeliers, creating a sophisticated sense of relaxed opulence. In honour of Florence and its traditions, every guest room and suite is named after noted Italian artists, benefactors and nobility.

Designed by Bottega Veneta Creative Director Tomas Maier, the Bottega Veneta suite is a unique and expansive environment created for those who appreciate quiet luxury, unparallelled service and a refined sensibility. The palette of the one-of-a-kind suite is serene and sophisticated, featuring muted neutrals that create a warm and relaxing atmosphere. The suite is furnished with an inspired mix of pieces from the Bottega Veneta furniture and home collection adding elegance and attention to the details.

The St. Regis Florence Iridium Spa Suites, allow guests to renew and refresh mind, body and spirit with a variety of spa treatments. The Iridium Spa Suites feature three luxurious and spacious spa suites complete with white and grey Florentine marble enriched with gold and silver walls and custom-designed mosaics. Two of the three suites feature circular chromo-therapy Jacuzzi tubs and double beds, perfect for couple massages and treatments.

The St. Regis Florence features what is expected to be one of the most magnificent restaurant and bar lounges in the city. Etichetta offers a truly innovative epicurean culinary experience combining a prestigious wine list with sophisticated culinary creations. The restaurant interiors feature picturesque glassed art ceilings dating back to the 19th century and spectacular hand-blown Murano glass chandelier.

The outdoor library terrace, with views of the Church and Piazza Ognissanti, is the perfect place to join the Italian ritual of "aperitivo" and admire the local's Bella Figura walking along the Lungarno.

Completion date: September, 2011
Location: Florence, Italy

Designer: Filippo Brunelleschi, Bottega Veneta
Photographer: The St. Regis Florence

完成时间：2011 年 9 月
项目地点：意大利，佛罗伦萨

设计师：菲利波·布鲁内莱斯基；宝缇嘉·温妮达
摄影师：佛罗伦萨瑞吉酒店

For private dining and functions, the hotel offers a majestic Ballroom on the mezzanine, or for smaller functions the intimate "Cantinetta", a 15th century brick vaulted cellar designed for those who love fine wines paired with original Tuccany flavours.

The St. Regis Florence's meeting and banquet rooms provide an exceptional setting for social and business events. Of the seven available spaces, each is comprised of a thousand matchless decorative details. A total of 550 square metres ensure successful gatherings. The impressive Salone delle Feste represents the preferred address for high profile board meetings and gala dinners, while smaller venues are tailored for company conferences.

1. The lobby features a unique selection of custom-designed furniture, paintings, frescos and crystal chandeliers.
2. The library lounge creates a sophisticated sense of relaxed opulence.

1. 大堂内装饰着定制家具、画作、壁画和水晶吊灯
2. 图书休息室十分适合休闲放松

佛罗伦萨瑞吉酒店拥有81间奢华的客房和19间套房，其中包括一间宝缇嘉·温妮达特别设计套房。酒店的客房和套房反映了佛伦伦萨丰富的艺术底蕴，以独立的手工雕刻金箔装饰为特色，拥有三种不同色调搭配——美第奇、佛罗伦萨人和文艺复兴。每个设计概念都拥有一系列的独特的定制家具、绘画、壁画和水晶灯，营造出一种精致的休闲感。为了向佛罗伦萨和它的传统致敬，每间客房和套房都采用了著名的意大利艺术家、赞助人和贵族命名。

宝缇嘉·温妮达套房由宝缇嘉·温妮达创意总监托马斯·迈尔设计，为崇尚奢华、完美服务和高雅的人士提供了独一无二的环境。这间独一无二的套房宁静而精致，柔和的色彩营造出温暖、放松的气氛。套房既有宝缇嘉·温妮达的专属家具，又有增添优雅和细节感的家居之选。

佛罗伦萨瑞吉酒店的铱星水疗套房让客人能够以各式水疗来洗涤自己身心。铱星水疗套房拥有三间奢华而宽敞的水疗套房，采用白灰大理石和金银两色墙壁以及定制马赛克进行装饰。其中的两间套房拥有圆形色彩疗法按摩浴缸和双人床，适合情侣按摩和理疗。

佛罗伦萨瑞吉酒店拥有城中最著名的餐厅酒吧。餐厅由三位米奇林大厨主掌，将为宾客提供精致的美酒佳肴。餐厅室内设计以如画的艺术玻璃天花板和华丽的手吹玻璃吊灯组成。

露天图书室平台俯瞰着教堂和万圣节广场，是饮用意大利餐前酒和欣赏当地美景的绝佳地点。酒店在中层楼处设置了一个宏伟的宴会厅，专供私人就餐和举办活动；对于小型的活动，"康蒂尼塔"酒窖为热爱美酒的人士提供了完美的处所。

佛罗伦萨瑞吉酒店的会议和宴会厅为社交和商业活动提供了非凡的环境。七个会议场所都由无数无与伦比的装饰细节组成。550平方米的面积保证了集会的顺利进行。节日大厅适合举办高端会议和晚宴，而较小地点则为公司会议量身定做。

1. The St. Regis Florence features what is expected to be one of the most magnificent restaurant and bar lounges in the city – Etichetta.
2. The hotel offers a 15th century brick vaulted cellar designed for those who love fine wines paired with original Tuccany flavours.
3. The restaurant interiors feature picturesque glassed art ceilings dating back to the 19th century and spectacular hand-blown Murano glass chandelier.

1. 佛罗伦萨瑞吉酒店拥有城中最华丽的餐厅和酒吧——埃迪齐达
2. 酒店建于15世纪的砖结构酒窖专为爱酒人士设计，具有托斯卡纳风情
3. 餐厅设计以19世纪的玻璃艺术天花板和璀璨的慕拉诺玻璃吊灯为特色

1. Interspersed with hand-decorated wooden columns, art and one-of-a-kind antiques, each 84-square-metre Royal Suite offers enjoyment and intrigue. The splendid atmosphere draws inspiration from the Court of the Medici, Florence wealthiest and most influential family between the 13th and 17th centuries.

2. Junior Suites feature wall frescos and 15th-century style artwork which are telling of the city's creative history.

1. 手工装饰的木柱、艺术品和独一无二的古董让皇室套房84平方米的空间充满了愉悦和吸引力；华丽的氛围从美第奇（13世纪和15世纪佛罗伦萨最富有、最具影响力的家族）宫廷中获得了灵感

2. 普通套房以壁画和15世纪风格的艺术品为特色，讲述了意大利充满创意性的历史之路

1. Deluxe Rooms are designed to feel welcoming, comfortable and elegant. Each room recalls the glory of the 15th century, Florentine, Medici and Renaissance style. The Florentine style luxurious rooms are adorned with coffered ceilings, frescoes and brocades.

2. The Florentine style room of Premium Deluxe Amo River view

3. The art of Renaissance and the art of hospitality combined in a masterpiece of exclusive living. The 200-square-metre Presidential Suite features a spacious balcony.

1. 豪华房的设计让人感觉宾至如归，舒适而优雅；每间客房都分别再现了15世纪风格、佛罗伦萨风格、美第奇王朝风格和文艺复兴风格的荣耀。佛罗伦萨风格的奢华客房装饰着方格天花板、壁画及锦缎

2. 佛罗伦萨风格的顶级河景豪华套房

3. 文艺复兴风格的艺术品打造了独一无二的生活空间；200平方米的总统套房拥有一个宽敞的阳台

SHERATON BANGALORE AT BRIGADE GATEWAY

班加罗尔Brigade Gateway喜来登酒店

With the rise of its IT industry and its influence in the global economy, Bangalore is not only a major international city but also a very connected city. This plus its flourishing fashion industry, which adds another layer to the meaning of being connected, have led the designers to the driving inspiration for the interior design of the Bangalore Sheraton – The Glamour of Haute Couture.

Haute Couture means fresh, cutting-edge, yet timeless design; and that is the overall mood of the property the designers want to achieve. They intend for everyone to feel fabulous when inside the Bangalore Sheraton – it's a place to see and be seen for discerning business travellers, sophisticated tourists, and trendsetters.

The ground floor public areas design is inspired by the newest LV shops opening around the world in terms of its simple blend of materials, forms and trademark symbols and logo.

Guests enter the hotel lobby through a fabulously high atrium featuring breathtaking high walls pattern panels with Indian inspired motifs. The music from the lobby lounge pulls you in further.

Looking ahead guests will see on their left a swooping platform rising out of the sand stone pavement with two inviting steps that lead to a unique grand staircase and water feature up to the meeting room level. This platform marks the formal arrival for guests attending functions and meetings above.

Each registration pod is designed with dark wood and metal trims to resemble a display case of a fashion boutique. But most people will have their eyes on the feature wall behind the pods which is an elegant wood sculpture wall that's dramatically lit from all around.

The lift lobby is a microcosm of the main lobby with an impressive ceiling feature that comes up from each lift door and through light play indicating the arrival of each lift, these custom-made panels, with their Indian inspired design, help creating unique design elements that serve as a starting point for guests to discover the rest of this magnificent tower.

Design of the ballroom level is meant to feel very Chanel, which is about dressing and accessorising women through whimsical yet restrained design that results in timeless glamour.

The prefunction area expands the water theme from the lobby. Instead of chandeliers, the ceiling is adorned with suspended

Completion date: 2011
Location: Bangalore, India
Designer: Dileonardo

Photographer: Warren Jagger, Ian Gibb
Area: 36,696m²

完成时间：2011 年
项目地点：印度，班加罗尔
设计师：迪里奥纳尔多

摄影师：沃伦·贾格尔；伊恩·吉伯
面积：36,696 平方米

wood soffits that have crystal beads of light in various sizes, resembling morning dew on a leaf, quietly hinting the glamour of events taking place here. Grand wood panel walls with custom-designed inset sconces dance up and down the walls in a variety of colours creating a dynamic atmosphere for lively conversations. The grand ballroom features the most unique ceiling design Inspired by women's jewellery, the impressive rings of custom pendant lights and soffits in various sizes redefine grandeur. As a complement to the rich ceiling, contemporary acoustic panelling on the wall allows for coexistence of grandeur and intimacy. This sophisticated quality is interpreted throughout meeting and function rooms to maintain a cohesive design.

The guestroom design is a tribute to Giorgio Armani's fashion. The guestroom environment is tranquil and sophisticated. The layout is pragmatic yet luxurious. The colour scheme is subtle yet memorable. The architectural features are bold yet relaxing.

The built-in headboard design consists of highly lacquered wood, tinted glass, and leather. It towers over the bed to accentuate the height of the room while achieving a high degree of scale and comfort with the padded leather panel and tinted glass above it.

The millwork design is inspired by fine men's stores which is about having a compartment for every piece of clothing and accessory and having a great place to dress up and look good.

The bathroom design maintains the same attitude as the rest of the room. A modern vanity mirror with built-in lighting on either side sets the mood for a very clean and modern look. The vanity counter's asymmetrical elevation with a built-in shelf for towels and other amenities creates a harmonic composition with the contemporary wash basin on the counter.

Bene got the inspiration from the works of Gianni Versace, a man who was never afraid to combine bold forms, colours and patterns in his designs to create sporty, yet luxurious clothing.

In terms of planning, the designers have integrated the restaurant's requirements with that of a VIP lounge to create a very synergetic space. Upon entering through the shared entry from the terrace, one immediately notices the variety of wine display all around with the wine screen in front spanning into to wrap around the wine bar in the lounge. Clearly visible to the right is the Pizza/Seafood Bar in the restaurant. The Italian gastronomic experience begins here.

The Pizza/Seafood Bar is designed to be the main feature of the restaurant where guests can see how pizzas are made from start to finish. The counter is finished with warm wood with colourful, back-lit glass inlay; its ceiling above continues the design as a soffit to highlight the area and to create a suitable seating condition for hotel guests dining alone.

The toilet is carefully shielded by the vanity counter while the tub and shower open up to the bathroom for maximum light and comfort. The glass partition next to the tub features glass-enclosed electrical operated wood blinds which provide privacy without the maintenance concerns.

The walls are paved with large sandstone tiles that continue up from the floor with small glass tile inlay in key areas that enhances the refinement of the room.

随着信息技术产业的崛起及其对全球经济的影响力的增加，班加罗尔已经不仅是一个国际化城市，而是一个极具连接感的城市。它的时装业的迅速发展，为设计师提供了班加罗尔喜来登酒店的设计灵感——高级时装的魅力诱惑。

高级时装意味着新潮、前卫而又经典的设计；这也是设计师想获得的酒店整体效果。他们希望每个人都能在班加罗尔喜来登酒店里享受到无与伦比的体验——它将是眼光独到的商务旅行者、富有经验的游客和时尚潮人的最佳住宿场所。

一层公共区的设计从最新的路易威登店面中获得了灵感，采用它简单的材料、造型和商标图案标识的结合方式，客人从高挑的中庭进入酒店大堂，中庭的壁板上装饰着印度图案。大堂吧的音乐将你带到了远方。

向上看，宾客们会在左侧看到一个从砂石路上升起的平台，两级台阶通往独特的大楼梯和水景设施，引领着人们到达会议室一层。这个平台标志着客人正式到达了上面的功能和会议空间。

每个登记台都采用了深色木材和金属边饰设计，宛如时装店的展示柜一样。大多数人会注意到登记台后方的装饰墙，优雅的木雕刻墙壁的四周都被点亮了。

电梯大堂是整个大堂的缩影。电梯门上方令人印象深刻的天花板和指示电梯到达的嵌入式灯板，这些定制面板的印度风情设计营造出独特的设计元素。客人们由此出发，前往探索大厦更多的景象。

宴会厅的设计极具香奈儿风格——通过异想天开而又拘谨的设计来为女士着装和搭配，形成经典的迷人魅力。

准备区延续了大堂的水景主题。天花板上没有使用吊灯，而是采用了悬吊的木拱腹和各种尺寸的水晶灯头，宛如叶子上的露珠一般，静静地展示着即将举办的活动的优雅魅力。定制的嵌入壁突式彩色烛台在宏大的木镶板墙壁上下跳动，为空间营造了动感的氛围。

大宴会厅拥有最特别的天花板设计。受到女士珠宝的启发，定制吊灯的吊环和各种尺寸的拱腹重新定义了华丽。作为天花板的组成元素之一，现代化隔音板保证了宏伟和私密的共存。会议室和活动室同样采用了这种精致的品质，形成了统一的设计。

客房设计向乔治·阿玛尼的时装设计表达了敬意。客房环境幽静而精致。整体布局实用而奢华。色彩搭配精巧而令人难忘。建筑特色大胆而放松。

1. Upon entering through the shared entry of Bene, one immediately notices the variety of wine display all around with the wine screen in front spanning into to wrap around the wine bar in the lounge. Clearly visible to the right is the Pizza/Seafood Bar in the restaurant.
2. Natural woods, granites and marble in a clean Californian style are set off by colourful accents at the background of bar counter.

1. 进入班尼餐吧的共享入口，人们马上会注意到酒架上展示的各种美酒。酒架延伸了休息室内整个吧台的空间。酒架的右边是比萨海鲜吧
2. 餐厅吧台的背景墙采用天然木质、花岗岩和大理石装饰，尽显别具一格的加州风情

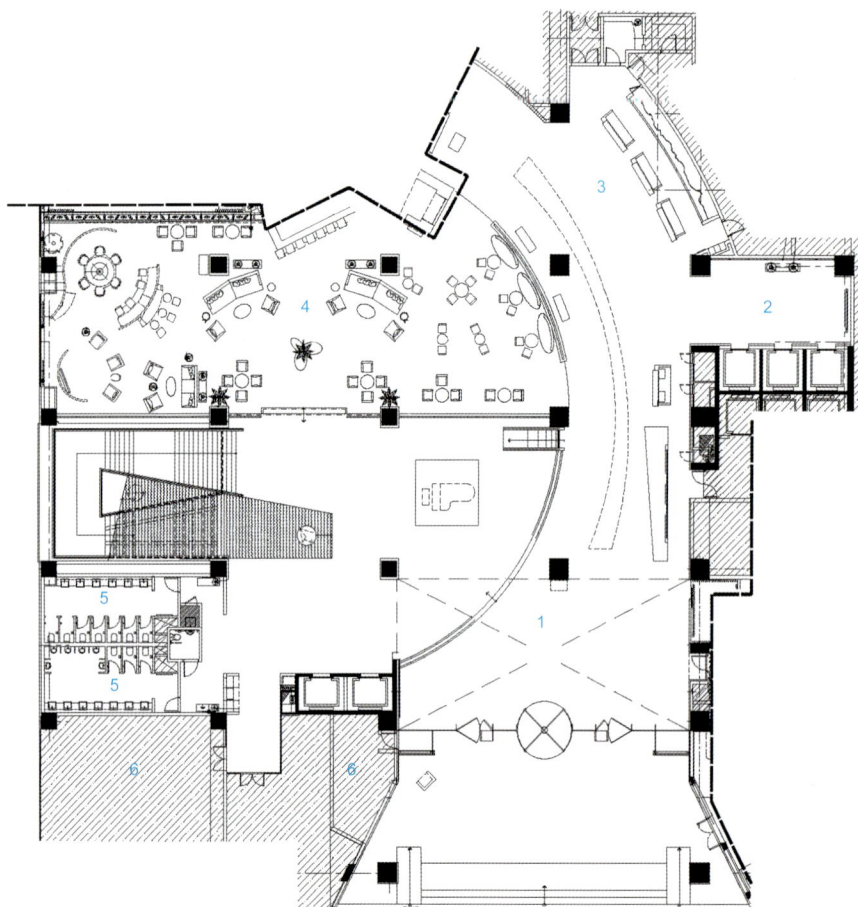

1. Entry
2. Lift lobby
3. Reception
4. Lobby lounge
5. Restaurant
6. Shop

1. 入口
2. 电梯大厅
3. 前台
4. 大堂酒廊
5. 餐厅
6. 商店

嵌入式传统设计由油漆木板、彩色玻璃和皮革组成。它盘踞在床头，凸显了房间的高度。同时，软垫皮革板和上方的彩色玻璃也突出了宏大感和舒适感。

木工设计从男装精品店中获得了灵感——精品店的每件服装和配饰都有其单独的隔间，为客人们提供了更衣和照镜子的好去处。

浴室设计与客房内的其他部分采用了同样的风格。现代的穿衣镜两侧设有照明，拥有简洁而现代的外观。梳妆台的对称正面设有嵌入式毛巾柜和其他设施，与现代的洗脸盆和谐地结合在一起。

抽水马桶与梳妆台被精心地隔离开，而浴缸和淋浴则拥有最大化的灯光和舒适度。浴缸边的玻璃隔断以玻璃密封的电子木制百叶窗为特色，既保证了隐私，又无需保养维护。

墙壁上的大块砂岩瓷砖延伸于地面，主要区域的玻璃砖嵌入物提升了空间的精致感。

班尼餐厅从詹尼·范思哲的作品中获得了灵感。范思哲从不畏惧大胆的造型、色彩和图案，他设计的服装花哨而富有奢华感。

在规划上，设计师结合了餐厅的基本要求，加入了一个非常协调的贵宾休息室。从平台进入共享入口，人们马上会注意到酒架上展示的各种美酒。酒架延伸了休息室内整个吧台的空间。酒架右边的比萨海鲜吧，将提供最正宗的意大利美食。

比萨海鲜吧的设计是餐厅的主要特色之一，客人们能看到比萨制作的全过程。柜台采用温暖的木材和彩色背光玻璃镶嵌物进行装饰；它的天花板形成了一个拱腹，凸显了整个区域并且营造了独自就餐的合适场所。

1. The lounge provides the comfortable environment and the necessary seat count for peak hours. The buffet counter is centrally positioned to be convenient for all guests.

2. In the FEAST, buffet counters and cooking stations are designed to resemble soft curves of rock formations. They are arranged to create an urban street market feel.

1. 休息室提供了舒适的环境，在高峰时期也有足够的座椅；自助餐台设在中央，方便宾客取用

2. 盛宴餐厅的自助餐台和烹饪台的设计模仿了柔和的岩山曲线；它们打造了城市街头市场的感觉

1. The grand ballroom features the most unique ceiling design. Inspired by women's jewellery, the impressive rings of custom pendant lights and soffits in various sizes redefine grandeur.

2. There are several design features in DurBar, from a beautiful custom-made glass chandelier to the glass towers as central element of the bar, moving to oversize sofa seating and complemented with wall paintings highlighting the grandeur of the Indian Royal Ancestors.

1. 大宴会厅独特的天花板设计从女士珠宝中获得了灵感，定制吊灯不同尺寸的圆环和内圈重新定义了奢华

2. 杜尔酒吧拥有美丽的定制玻璃吊灯和中央玻璃塔，宽大的沙发和壁画凸显了印度皇室风格

1. In addition to the luxurious modern furnishings, such as the banquette, lounge chairs, and sectional sofas, the lounge will have a Sheraton link area and a large back-lit wall as a backdrop that ties everything together.
2. The chaise lounge chair in the living room of Junior Suite offsets the formal qualities of the room layout for an added degree of casual elegance.
3. The built-in headboard design consists of highly lacquered wood, tinted glass, and leather.

1. 除了卡座、躺椅、组合沙发等现代奢华家具之外，休息室的喜来登连接区和大型背光墙将各个元素联系在一起
2. 普通套房各厅里的躺椅消减了房间布局的正式感，增添了轻松的氛围
3. 嵌入式床头由高光木板、彩色玻璃和皮革组成

SHERATON HUIZHOU BEACH RESORT

惠州金海湾喜来登度假酒店

Owning 300 guestrooms with mountain and sea view, and surrounded by a beautiful natural bay, Sheraton Huizhou Beach Resort is located in the famous resort area of Xunliao Bay in Huizhou, which is known as the "Oriental Hawaii".

The design of the hotel is pursuing distinct personality and humanity connotation. On the basis of internationalised style, using a modern eclectic methodology, it is relaxing and rustic, without losing the high standard of a five-star hotel. The designer and his team let the interior design extend and enrich the Chinese style of the overall architecture, to bring to the hotel a standing-out Chinese characteristic and oriental cultural verve.

Embellished by the featured Lingnan architecture style, the overall design is an ingenious marriage of oriental style and Western charm. The contemporary design is blended with the wooden truss, caisson, rockery and the traditional Lingnan cultural symbol. With its design composed of dots, lines and surfaces, the hotel lobby becomes a simple but chic integrity. Fish chasing on the giant sand wall presents the folk culture of Teochow fisherman. The water passage that connects the lounge bar with the exterior has introduced the exterior ocean view into the lobby and created a naturally harmonious space. The designer adopted stylish Italian furniture and decorated it with Chinese canework and wooden hollow-out, bringing a perfect combination of Chinese and Western styles. The visual effect of putting natural wildness into interior design, combined with the paradisiacal landscape, will reveal an international five-star resort hotel feature.

Completion date: June, 2008
Location: Huizhou, China
Designer: Bangsheng Yang

Photographer: Fang Jia, Yi Chen, Yongchang Wu
Area: 70,000m²

完成时间：2008 年 6 月
项目地点：中国，惠州
设计师：杨邦胜

摄影师：陈乙、贾方、吴永长
建筑面积：70,000 平方米

惠州金海湾喜来登度假酒店，位于有东方夏威夷美誉的惠州巽寮度假区，有山海景客房300余间，环拥美轮美奂的金海湾自然生态海湾。

该项目的室内设计沿袭建筑的中式风格，在打造中国特色和东方文化神韵的同时，凸显鲜明的个性和人文内涵。运用折中的现代中式手法，休闲质朴但不失其国际化五星级品质。

整体设计融东方情调与西方韵味于一体，巧妙将中国岭南建筑特色的木构架、藻井、假山叠石等元素融入现代设计中，并与客家民居的白墙、灰砖、潮州木雕、天井的传统岭南文化符号相结合。大堂设计简约时尚，空间由点线面构成，线的广东居民坡屋顶，面的大面积实墙，简单过渡，连之一体。巨幅沙雕背景，以群鱼追逐彰显潮州渔民民俗风情，古朴而逸趣。水面浮道，连接大堂吧和室外，水面与海面相连，海天一色，静谧由室内向户外延伸。室内陈设选取意大利现代潮流家具，饰以中式藤编、木质镂空，一派中西浑然天成，纵深贯穿酒店。时空错位的视觉感受，山海相邻的非凡体验，尽显国际五星度假酒店的休闲风范。

1. With its design composed of dots, lines and surfaces, the hotel lobby becomes a simple but chic integrity.
2. The lounge in the lobby features custom-design furnishings, creating tranquil atmosphere.

1. 点线面的设计让酒店大堂显得简洁而时尚
2. 大堂休息室采用了定制装饰，营造出静谧的氛围

1. Embellished by the featured Lingnan architecture style, the overall design is an ingenious marriage of oriental style and Western charm.

2. Under Lingnan architecture style ceiling, sitting casually in the cane chairs, you can enjoy the cosy atmosphere of Pool Side Bar.

3. The designer adopted stylish Italian furniture and decorated INAZIA – Asian restaurant with Chinese cane work and wooden hollow-out.

1. 整体设计采用了岭南建筑风格，结合了东西方设计风格
2. 在岭南建筑风格的屋顶下，随意坐在藤椅中享受池畔酒吧的惬意氛围
3. 设计师采用了时尚的意大利家具，并在亚洲餐厅里添加了中式藤条和木格装饰

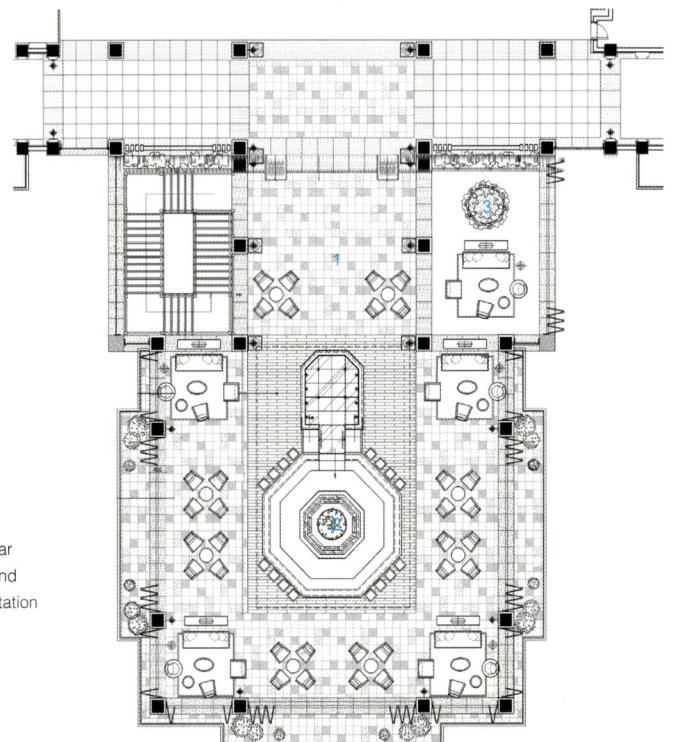

1. Lobby bar
2. Bandstand
3. Flower station

1. 大堂吧
2. 演奏台
3. 花台

1. The interior design in the dining area is combined with the paradisiacal landscape.

2. The water passage has introduced the exterior ocean view into FEAST.

3. The guestrooms are spacious and stylishly designed with your utmost comfort in mind.

1. 就餐区的设计与绝美的景观相互结合
2. 水道将外面的海景引入了宴会厅
3. 客房宽敞而时尚，极其舒适

THE PENINSULA HOTELS GROUP
A Historical Legend of the Bustling Times

半岛酒店集团——喧嚣时代的历史传奇

THE PENINSULA
HOTELS

As the first hotel of The Peninsula Hotels Group, The Peninsula Hong Kong is founded in 1928, with a history of more than 80 years. Currently, the Peninsula Hotels Group only owns 9 hotels all around the world. The group targets at social celebrities and high-class business journey, providing extraordinary experience for guests. The hotels' architectural and interior design is the most amazing part. With architectural and interior design standards set by the Grande Dame – The Peninsula Hong Kong, the unified design style of the Peninsula Hotels provides guests with a deep and reserved sense. The hotels integrate oriental culture gradually into the commanding occident design concepts and insert the historical and artistic gentle charms into the contemporary bustling cities.

The overall architectural design style of the Peninsula Hotel Hong Kong roots in its history. During the Republican period, the afternoon tea of the Peninsula Hong Kong is the design inspiration of Eileen Zhang's Love in a Fallen City. After World War II, the first governor of Hong Kong signed the surrender here; afterwards, Queen Elizabeth II and the former US president Ronald Reagan appointed the Peninsula Hong Kong as their exclusive hotel to stay. During the more than 80 years, the Peninsula Hotel Hong Kong has always been the place for various celebrities and nobilities to stay and enjoy lifestyles. This built up the deep historical contents and therefore influenced the overall architectural design style. Different from other modern skyscraper hotels, without unique façades nor new building materials, the 9 Peninsula Hotels use the most ordinary materials to express a decent and reserved effect, which also remains the historic sense. For instance, the Peninsula Hong Kong remains the architectural style of the 1930s and although completed in 2009, the Peninsula Shanghai restores the architectural style of Shanghai's "golden era".

The interior design style of the Peninsula Hotels still implies a deep historical atmosphere. Most of the Peninsula Hotels use Art Deco style in their interior design, which represents the review from modern industrial culture to historical culture, emphasise symmetrical spaces and take metal colours as their main decoration. The elegant geometric patterns are often used in the decorations, such as the symmetrical layout of the bathroom, the winding lines on the wall, the cushions with diamond patterns on the bed or the sofa and Egyptian relief on the columns in the lobby. Furthermore, to be noticed, the Peninsula Hotels combine modern occident design methods and oriental culture harmoniously, without any conflicts caused by oriental elements. The Peninsula Tokyo is a classic combination of oriental and occident cultures. On one hand, the hotel invited 60 Japanese artists and artisans to create various decorations; on the other hand, Tino Kwan's lighting design is nearly perfect. As we all know, lighting design will enhance the overall contemporary effect. The lighting design in the Peninsula Tokyo plays more than that. For example, when guests step up in the spiral staircase, the lighting bridge with glass patterns will present a curved effect to enhance the Japanese local design style.

The Peninsula Hotels Group stands in the bustling cities with its unique gentle and cultural historic charms. This chapter will display the architectural and interior designs of the Peninsula Hong Kong, the Peninsula Tokyo and the Peninsula Bangkok respectively, revealing a legendary history of the Peninsula Hotels.

作为半岛酒店集团的第一家酒店，香港半岛酒店成立于 1928 年，至今已有 80 余年的发展史，在全球拥有 9 家本品牌酒店。集团专门针对社会知名人士及高级商旅，为其提供最非凡的体验感受，而其中尤以酒店建筑和室内设计最为惊艳。细数 9 家半岛酒店，在以有"东方贵妇"之称香港半岛酒店的建筑与室内设计标准之下，酒店的设计风格给客人以深沉和内敛的之感，它将东方文化潜移默化地融入西方主流的设计理念，更将富于历史与艺术韵味的温文尔雅嵌入喧嚣的时代都市。

半岛酒店这种整体的建筑设计风格根植于本身的历史传奇。民国时期，正是半岛酒店的下午茶给予了民国才女张爱玲谱写《倾城之恋》的灵感源泉；第二次世界大战结束后第一任港督杨慕琦正是在此签署投降书；此后，英女皇伊莉莎白二世，美国前总统里根都指定香港半岛酒店为唯一下榻酒店。经历 80 余年的时代变迁，半岛酒店一直是各都市名流杯影交错，贵族浪漫挥霍之所。这使半岛酒店积累下深厚的历史底蕴，随之影响其整体的建筑设计风格。纵观 9 家半岛酒店的建筑设计，与现代高耸入云的摩天酒店建筑有些不同。它们没有奇特的外立面，更不用那些新式的建筑材料，只是用最普通的材料传达出一种庄重深沉的效果，并且保留了一份历史的痕迹感。例如香港半岛酒店的建筑设计，保留了 20 世纪 30 年代建筑风格，而上海半岛酒店，虽然在 2009 年落成，却还原了上海"黄金时代"的建筑风貌。

半岛酒店的室内设计风格依然隐含着深沉的历史气息。多数半岛酒店在室内装饰中运用 Art Deco 装饰风格，Art Deco 装饰风格本身代表着一种现代工业文化向历史文化的追溯，强调空间的对称，习惯使用金属色为装饰主色，优雅的几何图形经常被运用到装饰物之中。例如浴室对称的布局，墙面点缀回纹饰曲折线条，床上或沙发上放着有菱形压纹装饰的靠垫，大厅的立柱装饰着埃及神话的浮雕。

另外，值得称道的是，半岛酒店将现代西方的设计手法与东方文化相融合并迸发出精彩的火花，而又不因为这些东方元素而产生冲突感。例如东京半岛酒店，一方面酒店邀请近 60 位日本本土艺术家、工艺匠设计并打造酒店内的各式装饰品；一方面酒店的灯光设计师关永权的巧妙设计也堪称完美。众所周知，优秀的室内灯光设计会令室内整体现代感十足。而在东京半岛酒店，灯光的设计效果却不止这些，例如当客人随一盘旋而上时，楼梯旁配以玻璃图案的灯桥呈现弧形效果，反而增强了日本本土设计风格。

在当今的充满喧嚣的繁华都市，半岛酒店集团以它温文尔雅的历史怀旧气息屹立之中。本章将逐一展示香港半岛酒店、东京半岛酒店、曼谷半岛酒店的建筑与室内设计风貌，逐一破解半岛酒店的历史传奇。

THE PENINSULA HONG KONG

香港半岛酒店

Located on Salisbury Road, Tsimshatsui, overlooking Victoria Harbour, The Peninsula is right at the heart of Kowloon's business, shopping and entertainment district. The Peninsula's premier suite, The Peninsula Suite, is located on the 26th floor of the tower, and offers superb views of the harbour in the most spacious and luxurious setting. In the original building, five guest floors contain a total of 141 guest rooms and 27 suites, while the tower features 11 guest room floors comprising 105 guest rooms and 27 suites.

The tower benefits from the raising of the former height restriction on the Kowloon peninsula, and thus all tower rooms and suites have wholly unobstructed views of either Hong Kong Island and Victoria Harbour to the south, or the city and mountain ranges of Kowloon to the north.

Among the most spacious in Hong Kong, all rooms and suites are furnished to the very highest standards of luxury in a fresh classic European style. Subtle oriental influences are reflected in the rich fabrics which follow a varying theme of deep blue or green, gold and ivory.

The Peninsula Spa offers a deeply personalised spa experience like nowhere else in Hong Kong. The 1,116 sqm Spa occupies two floors, offering sweeping views of Victoria Harbour and 14 state-of-the-art treatment rooms, making for an oasis of calm in the city for guests to renew themselves in mind, body and spirit.

The Peninsula Fitness Centre comprises a pool and a health club. The Peninsula's swimming pool (7th floor) has been designed to offer swimmers uninterrupted views of the Hong Kong skyline, whilst a retractable glass screen can be closed over the front elevation of the entire pool area to allow year-round use. The pool is 18 m long, 1.5 m deep, and is modelled on a classical Roman theme with columns, friezes, statues and niches to evince a palatial setting. The pool is linked to a sun-terrace and Fitness Centre, which extends out over the west wing of the original building and covers an area of approximately 1,600 sqm.

Restaurant

The Lobby is the most elegant meeting place in Hong Kong. With marble-topped tables, sumptuous seating and exclusive Tiffany chinaware, the Lobby makes for a perfect all-day dining venue from breakfast onwards.

Arched sea-facing windows and ceiling fans circling lazily above crispy white linen-covered table amid elegant palm fronds, the Verandah is elegant yet casual with a light and airy ambience. The décor feature beige and ivory, hanging fixtures, rattan upholstered

Completion date: 2008 (renovation)
Location: Hong Kong, China
Designer: Richmond International, Allison Henry Designs
(Public Area, Guest Rooms/Suites and Function Rooms),

Rocco Design Limited (Building Architect)
Photographer: The Peninsula Hong Kong
Area: 100,000m²

完成时间：2008（翻新）
项目地点：中国，香港
设计师：里士满国际；阿里森·亨利设计（公共

区域、客房/套房和活动场地）
许李严建筑师有限公司（建筑）
摄影师：香港半岛酒店
面积：100,000平方米

chairs, wooden and marble floors and fluted columns. The table display The Peninsula's exclusive Tiffany chinaware and this could traditional silverware. The restaurant's Mediterranean cuisine places special emphasis on light and healthful fare, yet for those with a sweet tooth there is a tempting choice of desserts.

Located on the 28th floor of The Peninsula tower, Felix is the creation of the renowned avant-garde designer Philippe Starck. The restaurant offers breathtaking views of Victoria Harbour, Hong Kong Island and Kowloon. Other features include the Wine Bar, the Balcony, the American Bar and the Crazy Box, a small discotheque.

1. Main dining room in Gaddi's
2. Impressing wood interior and oriental rugs with sliver ornaments and sepia photos line the walls, the traditional Cantonese restaurant – Spring Moon blends art décor into the classic Shanghainese design style.
3. French dining restaurant – Gaddi's is decorated with shimmering candlelight and rich blue and gold carpet, reflecting European décor influence.

1. 吉地士餐厅
2. 嘉麟楼中餐厅结合了装饰艺术和古典粤式设计风格，采用木材装饰和东方地毯，配有银色装饰和棕褐色照片来装饰墙壁
3. 吉地士法式餐厅采用了闪烁的烛光和丰富的蓝色和金色地毯进行装饰，反映了欧式风格

香港半岛酒店坐落在尖沙咀梳士巴利路，俯瞰维多利亚湾的美景，处在九龙商业、购物和娱乐区的中心。半岛酒店的顶级套房——半岛套房坐落在塔楼的26楼，客人将在这个舒适而奢华的环境中俯瞰港口的壮丽景色。在原有大厦中，5层客房楼层共有141间客房和27间套房，而塔楼的11层客房楼层则拥有105间客房和27间套房。

塔楼受益于九龙岛楼高限制的提升，让所有客房和套房都能享受香港岛、维多利亚港、城市和九龙山的景色。

作为香港最宽敞的客房和套房，所有房间都采用极致奢华的经典欧式主义装饰。丰富的纹理同样反映了精妙的东方特色，以深蓝色、绿色、金色和象牙色为主题。

半岛水疗中心将提供香港独一无二的深层私人水疗体验。1,116平方米的水疗中心占据两个楼层，坐拥维多利亚港的优美景致，拥有14间先进的理疗室，是一座城市绿洲。客人可以在这里让身心得到舒展。

半岛健身中心由游泳池和健康俱乐部组成。半岛游泳池（位于8楼）为游泳者提供了香港天际线的美景，可伸缩式玻璃屏幕让整个泳池区域在全年都能够使用。泳池长18米，深1.5米，采用古典罗马主题：廊柱、檐壁雕像和壁龛共同营造出富丽堂皇的背景。泳池与阳台和健身中心相连。一直延伸到原始建筑西翼

的健身中心，总面积约1,600平方米。

餐厅

大堂茶座可谓是香港最优雅的会面场所。大理石台面、豪华的休息区以及奢侈的蒂凡尼瓷器让大堂茶座成为完美的全天候就餐场所。

吊式电扇在雪白的桌面上慵懒地转动，餐桌四周环绕着优雅的棕榈叶，配合着宽大的海景拱窗，露台餐厅优雅而不失休闲，整体气氛轻松愉悦。室内装饰以淡黄色和象牙色为主，布满了悬垂灯具、藤条椅、木地板、大理石地面以及廊柱。餐桌上展示着半岛酒店的独家蒂凡尼瓷器和传统银器。餐厅的地中海美食特别注重轻食主义和健康饮食，同时又为喜好甜点的人提供了各种精美的点心。

位于28楼的Felix餐厅由知名先锋设计师菲利普·斯塔克一手打造。餐厅享有维多利亚港、香港岛和九龙的美景。餐厅还包括红酒吧、包厢、美国吧以及小型迪斯科舞厅疯狂盒子。

1. The private dining room in Gaddi's
2. The classic elegance of the Salisbury Room is reflected in the massive mirrors on the wall.

1. 吉地士餐厅的包房
2. 利士厅的经典高雅体现在墙上的大镜子上

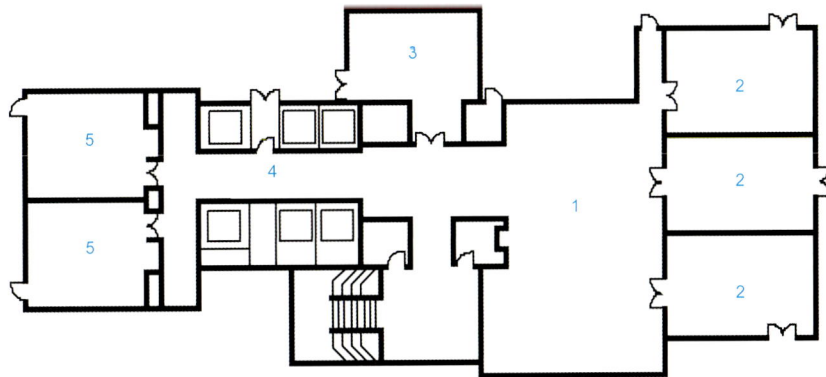

1. Salisbury Foyer
2. Salisbury Room
3. The Golden Pen
4. Guest lifts
5. Nathan Room

1. 利士门厅
2. 利士厅
3. 金笔厅
4. 客梯
5. 弥敦厅

1. The plush décor in deluxe suite makes its own elegancy.
2. Deluxe Harbour View Suite resembles a vintage world with its deep pile carpets, a faux fireplace, oriental floor rugs and high wing-backed armchair.
3. Grand Deluxe Kowloon View Room features the unique Peninsula style.

1. 豪华套房的奢华装饰营造了一种独特的优雅
2. 特级豪华海景套房利用长毛绒地毯、假壁炉、东方地毯和高背扶手椅来打造了一个复古的世界
3. 特级豪华九龙景观客房拥有独特的半岛酒店风格

THE PENINSULA TOKYO

东京半岛酒店

The combined work of celebrated architect Kazukiyo Sato and interior designer Yukio Hashimoto brings a new aesthetic to Japan with the creation of The Peninsula Tokyo, a new landmark hotel where Japanese culture and seasons are reflected in a thoroughly modern context throughout the hotel's design and facilities.

While other luxury hotels have opted for international décor for their properties located within commercial buildings or multi-purpose complexes, The Peninsula Hotels deliberately selected a modern Japanese ambience for its latest property. The only free-standing hotel to be opened in Tokyo in more than a decade, The Peninsula Tokyo is contemporary, yet echoes Japan's rich heritage and culture in its design — both exterior and interior — including a plethora of Japanese artworks totaling nearly 1,000 and created by nearly 60 artists, 90% of them Japanese using traditional techniques and methods.

Kazukiyo Sato's vision of the hotel as a traditional Japanese lantern standing proud at the entrance to Marunouchi and Ginza sets the tone from the outset. The exterior is amber Namibian granite, which provides a pleasant contrast to the neighbouring grey stone buildings, while the hotel's forecourt features a raised fountain of a ji-ishi Japanese granite stone from Kagawa Prefecture with a landscaped garden of classical pine, cherry and maple trees, reflecting the Japanese love of nature and the changing seasons. Large plants located at each side of the main entrance change according to seasons.

Two storeys high, the signature Peninsula Lobby features ivory walls with wooden lattices, echoing the senbongoshi of old Kyoto, the nation's former capital and still the centre of Japanese culture. The lattice motif is also found throughout the hotel, in corridors, guestrooms and public areas.

Guest floor corridors reference the streets of old Kyoto, with andon floor lights and granite and mirror panels set into the walls resembling water, while the carpets feature a woven kimono thread pattern. The slanted metal room number panels beside each door take an origami theme, with backlights shining through acrylic washi paper panels as daylight shines through folded origami paper.

At the foot of the grand staircase is a traditional Japanese karensansui rock garden, with white sandstone pebbles "swirling" around granite rocks, depicting variously water, islands and the universe.

Completion date: 2007
Location: Tokyo, Japan
Designer: Kazukiyo Sato, Yukio Hashimoto

Photographer: The Peninsula Tokyo
Area: 43,000m²

完成时间：2007 年
项目地点：日本，东京
设计师：佐藤和清、桥本夕季夫

摄影师：东京半岛酒店
面积：43,000 平方米

1. With marble-topped tables, sumptuous seating and exclusive chinaware, the Lobby is the most elegant meeting place in Tokyo.
2. Decorated with the unique ornament under the ceiling, guests can experience the avant-garde design in Peter private dining room.

1. 两层楼高的半岛大堂采用了象牙色墙面和木制栅格，与京都的传统栅格遥相呼应
2. 拥有独特天花板装饰的彼得餐厅包房让宾客体验了先锋设计

由著名建筑师佐藤和清与室内设计师桥本夕季夫联手打造的东京半岛酒店为日本带来了全新的美学意识。这座地标性酒店的设计和设施反映了现代环境下的日本文化和四季变化。

许多奢华酒店都选择将地址选在商业建筑或多功能综合体之中，配以国际化装潢设计。半岛酒店另辟蹊径，选择打造自己的现代日式风格。作为东京近十年来唯一的一座独立式酒店，现代的东京半岛酒店在室内外设计上都反映了日本丰富的遗产和文化，它拥有由近60位艺术家所打造的近千件日本艺术品，其中的90%都采用了日本传统工艺。

佐藤和清将酒店设计成一个耸立在丸之内和银座门户的传统日式灯笼。酒店外部采用琥珀色纳米尼亚花岗岩，与周边的灰色石造建筑形成了愉悦的对比。酒店前院的喷水池采用了来自香川县的传统日本花岗岩，辅以由松树、樱花和枫树组成的景观花园，反映了日本人对自然的热爱和四季的变换。主入口两侧的大型景观植物会随着季节的变换而变换。

酒店大堂采用了象牙色墙面和木制栅格，与京都的传统栅格遥相呼应。同时，栅格图案在酒店的走廊、客房和公共区域随处可见。

客房层的走廊参考了京都的街道，采用安藤地灯与嵌在墙上的大理石和镜子来模仿水面，而地毯则呈现出和服走线图案。客房门边倾斜的金属门牌采用了折纸主题，背景灯透过亚克力和纸板而闪耀，仿佛日光穿透折纸一样。

大楼梯的脚下是一个传统日式山水岩石花园，纯白的砂石卵石环绕着花岗岩，描绘出山水、小岛和整个宇宙。

1. Wedding set-up in The Ginza Ballroom
2. Hei Fung Terrace Private Dining Room inherits the design theme from the Peninsula Hong Kong's Spring Moon.
3. Setting with vaulted ceiling and natural light, the chapel can be the most romantic wedding venue.

1. 银座宴会厅婚礼布置
2. 凤凰台私人包房传承了香港半岛酒店嘉麟楼的设计主题
3. 拱顶和自然光线让小礼拜堂成为最浪漫的婚礼会场

3

1. Traditional Chinese window lattice setting in Hei Fung Terrace tells the origin of the hotel.
2. Guests can be served in the traditional Suzhou garden-like restaurant – Hei Fung Terrrace.

1. 凤凰台的传统中式窗格彰显了酒店的传统
2. 宾客可以在苏州园林式餐厅——凤凰台——就餐

2

1. Japanese Ceremony Room offers the traditional atmosphere, featuring a coved washi paper ceiling, an urushi striped wall behind the altar and shoji sliding doors.

2. Colours and textures used in the Grand Ballroom are reminiscent of those used in kimonos with the carpet pattern that of orimano, or randomly woven kimono threads, with shoji screen-like lighting.

3. With high ceilings and pillarless interiors, the Ginza Ballroom is designed for medium-size meetings.

1. 日式典礼厅具有传统氛围,以内凹和纸天花板、生漆条纹墙和障子拉门为特色

2. 东厅适合举办中等规模的会议,采用了挑高天花板和无柱式设计

3. 大宴会厅在色彩和纹理上与和服设计相连:地毯采用了和服走线图案,而障子式的灯具则采用了和服纹理

THE PENINSULA

1. The same amber Namibian granite as the hotel's exterior cladding is used, while the oval shape of the ceiling reflecting in the water of the pool below resembles the moon.

2. Traditional elements are found in suites, with the carpet design of karamatsu pine and tsura (crane) or linen leaf pattern on the table lamps. As the linen plant grows very rapidly, the motif denotes fast growth and prosperity.

3. Guests in Deluxe Park View Room can appreciate the design of Yukio Hashimoto: rich earth tones with woods and lacquer touches.

1. 游泳池采用了与酒店外观一样的纳米尼亚花岗岩，天花板的椭圆形造型倒映在水面上，像一轮明月

2. 套房体现了多个传统元素：地毯上的松鹤图案和台灯上的亚麻叶图案；由于亚麻生长十分迅速，图案暗示着快速生长和繁荣兴盛

3. 客人们可以在豪华园景客房欣赏设计师桥本夕纪夫利用木材及漆器陈设元素营造出的室内布局

1. Among the largest in Tokyo, rooms and suites bond nature's elements with traditional Japanese skills to provide a tranquil and serene setting.
2. The unique location of the hotel gives the designer chance to create Deluxe Park View Room with special visual impact, both inside and looking outwards.
3. The spectacular polished and rough-hewn granite and cherry wood bathrooms complete with large soaking tub and stone faucet offer Japanese hot spring ambience.

1. 半岛酒店拥有东京最大的客房和套房，其设计融合了自然元素和传统日本工艺，打造了宁静平和的氛围
2. 酒店独特的地理位置让设计师能够为高级公园套房打造由内至外的独特视觉效果
3. 浴室采用抛光和磨砂大理石和樱桃木进行装饰，配有巨大的浴缸和石制水龙头，营造出日式温泉的氛围

1. Living room
2. Bedroom
3. Foyer
4. Dressing room
5. Bathroom
6. Entrance
7. Powder room
8. Connection with Grand Deluxe Room

1. 客厅
2. 卧室
3. 门厅
4. 更衣室
5. 浴室
6. 入口
7. 化妆间
8. 与豪华房相连

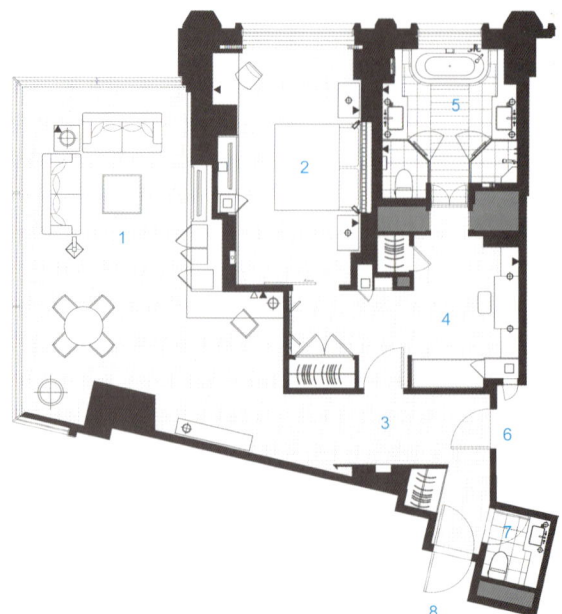

THE PENINSULA BANGKOK

曼谷半岛酒店

One of the finest hotels in Asia, The Peninsula Bangkok, perfectly mixes timeless elegance and sophistication with contemporary comfort and technology. Its distinctive waved-shape design affords each of the 370 rooms and suites a stunning panoramic view of the historic Chao Phraya River.

Traditional but luxurious guestrooms have separate sleeping and living areas and magnificent marble bathrooms. With unrivalled hotel facilities, The Peninsula Bangkok features the magnificent Peninsula Spa, offering treatments that layer Thai, European and Ayurvedic philosophies for an entirely original experience. Guests can relax in a Thai pavilion by the 88-metre three-tiered swimming pool, pamper themselves at the beauty salon or indulge in the Fitness Centre's gymnasium, sauna, whirlpool and steam rooms.

All of the hotel's outlets are ideal for relaxing and unwinding, including the Bar, the Pool Bar and the River Bar. At the tranquil riverside retreat of The Peninsula Bangkok, guests can be certain that they will receive legendary Thai service combined with Peninsula perfection.

Completion date: November, 1998
Location: Bangkok, Thailand

Designer: Glen Texeira Inc, Los Angeles; Denton
Corker Marshall Limited Hong Kong
Photographer: The Peninsula Bangkok

完成时间：1998 年 11 月
项目地点：泰国，曼谷

设计师：格伦·特谢拉公司；登顿·库克·马歇
尔公司香港分公司
摄影师：曼谷半岛酒店

1. Guests sitting in the River Bar can enjoy spectacular views of the River of Kings.
2. River Café and Terrace offers a relaxed experience in cool air-conditioned comfort.
3. Known for its salas with steep roofs and open sides, Thiptara offers the best dining experience and features home-style Thai cooking.

1. 客人可以在河畔吧欣赏河流的壮丽景色
2. 河边咖啡厅提供了凉爽舒适的体验
3. 泰普塔拉餐厅以斜屋顶开放式大厅而著称，提供了最好的泰国私房菜

作为亚洲最出色的酒店之一，曼谷半岛酒店完美地结合了经典的优雅和纯熟的现代技术。独特的波浪形建筑中的370间客房和套房都享有古老的湄南河全景。

传统而奢华的客房拥有独立睡眠区、起居区以及堂皇的大理石浴室。

在曼谷半岛酒店无与伦比的酒店设施中，半岛水疗中心尤为突出。水疗中心结合了泰式、欧式和印式理疗，打造了全新的健康理念。客人可以在88米高的三层游泳池旁的泰式亭中休憩，在美容沙龙中充分享受或是沉浸在拥有健身房、桑拿房、漩涡泳池和蒸汽浴室的健身中心里。

酒吧、泳池吧和河流吧一起将酒店打造成轻松愉快的理想之所。在曼谷半岛酒店宁静的河畔，客人们将尽享传奇的泰式服务和半岛酒店的特色经典。

1. The serene and elegant setting of Mei Jiang Chinese Restaurant
2. Unwind with refreshing cocktails at the Bar
3. Wedding set-up in Sakuntala Ballroom
4. Chinese wedding set-up in Chintara Room
5. Classroom set-up in Sakuntala Ballroom

1. 湄江中餐厅安静又优雅的室内布局
2. 在酒吧客人可以品尝爽口的鸡尾酒放松身心
3. 萨昆塔拉厅的婚礼布置
4. 乾塔拉厅的中式婚礼布置
5. 萨昆塔拉厅的会议布置

1. With teak floors and rich decorations made from Thai silk, the Grand Deluxe Suite features distinctive Thai style.
2. The Peninsula Suite offers unrivalled luxury with a unique panoramic view and comfortable atmosphere.
3. The elegance and comfort of textured silk fabrics and modern Thai furniture channels the interiors of traditional Bangkok houses. The Thai Suite features classic Thai luxury, from the private dining room chairs covered in gold silk to the sumptuous sofas.

1. 柚木地板和丰富的泰国丝绸装饰让高级套房享有独一无二的泰式风格
2. 半岛套房无与伦比的奢华体现在独特的全景视野和舒适的氛围上
3. 泰国套间的设计灵感源自传统泰式房屋，优雅舒适的泰丝布料陈设及充满时代感的泰式家具，私人餐厅座椅的金色丝绸椅套及华丽的沙发，古典泰式风情油然而生

1. Living room
2. Powder room
3. Foyer
4. Entrance
5. Dressing area
6. Bathroom
7. Bedroom

1. 客厅
2. 化妆间
3. 门厅
4. 入口
5. 更衣区
6. 浴室
7. 卧室

1

2

1. The terrace suite is upholstered in rich Thai silk, decorated with ceramic art pieces and furnished with Chinese antique furniture, creating a luxurious blend of Thai and Chinese style.

2. The Grand Deluxe Suite boasts distinctive Thai touches such as gleaming teak floors, rich silk furnishings and Asian art pieces.

1. 露台套房到处装饰着丰富的泰国丝绸，摆放着陶瓷艺术品和中式古董家具；奢华的氛围结合了泰国和中国的风格

2. 特级豪华套间享有油亮的柚木地板及亚洲工艺品摆设营造一缕泰式风情

1. Private GYM	7. Private study	13. Formal dining room
2. Master bathroom	8. Private living room	14. Terrace
3. Dressing area	9. Grand gallery	15. Guest bathroom
4. Anteroom	10. Entrance	16. Hallway
5. Powder room	11. Foyer	17. Guest bedroom
6. Master bedroom	12. Coat room	18. Sitting room

1. 私人健身房	7. 私人书房	13. 正式餐厅
2. 主浴室	8. 私人客厅	14. 平台
3. 更衣区	9. 大走廊	15. 客用浴室
4. 会客室	10. 入口	16. 走廊
5. 化妆间	11. 门厅	17. 客卧室
6. 主卧室	12. 衣帽间	18. 休息室

INTERCONTINENTAL HOTELS GROUP
Growing with the World

洲际酒店集团——同世界一起成长

INTER·CONTINENTAL®
HOTELS AND RESORTS

Founded in 1777, InterContinental Hotels Group has a history of more than 200 years. The group has more guest rooms than any other hotel company in the world – that's over 153 million guests annually, more than 658,000 rooms in over 4,480 hotels in nearly 100 countries and territories around the world. These large numbers reflect IHG's strong influences in global hotel industry. On one hand, the hotel's development depends on the push of globalisation; on the other hand, the growth of IHG is an epitome of global development and the design core of IHG exactly meets the need of global development.

Nowadays, the demand of sustainable environment is gradually increasing. According to this, IHG realises that hotels are facing an important issue of how to achieve the sustainable development and gradually transforms the focus of hotel design. The director of ASID Karin Verzariu has summarised the design of InterContinental hotels: the hotel emphasises environmental consciousness and sacrifices aesthetics to some extent. IHG introduces environmental awareness into hotel design and requires the designers to consider both the environmental influences of the hotel and the elegance and comfort. The conflict between the two requirements is that the design should pay more attention to its environmental influence, even to future environmental influence. An InterContinental hotel is not only visually attractive but also sustainable, avoiding any conflict with eco environment. In practice, the hotel promotes green materials and tries to reduce energy consumption in construction to protect environment and reduce pollution.

Today is an era of fast development and expression of individuality. The traditional luxurious business hotel no longer meets the requirements of business travellers. The pursuit of fashion and art spreads into various fields and as a fresh concept, design hotel immediately attracts people's attention. IHG, too, begins to pay attention to fashion and artistic expression of hotel design. The designs of InterContinental hotels intentionally create a pleasant feeling visually and mentally. Especially the decoration is both innovative and creative. In the design of InterContinental Puxi, LTW created a glass sculpture for the lobby. The spectacular glass sculpture on the ceiling is inspired by the Chinese ribbon dance, guiding guests into the reception naturally. In an InterContinental hotel, guests could pursue chic design. From decorative colours to furniture shapes, IHG possesses excellent sights and leads in fashion.

Now the world is undergoing a phenomenon of great cultural integration; the economic trade and information communication are developing too, which enable the hotel to become a gathering place for conveying and reflecting culture. As a representative for inheriting culture, IHG combines various cultural elements in its design. People can find some Chinese landscape elements, even some poetic images of garden in a high-end InterContinental hotel with modern facilities. The design uses borrowed scenery to add depths to the focus and levels and introduces natural elements into the interior, combining interior and exterior spaces skillfully and creating a harmonious space.

Apparently, IHG develops with the world's pace. In an InterContinental hotel, instead of unchanged standards and restraints of architectural and interior designs, we can feel the constant improvements in design requirement. This chapter selects three brand hotels of IHG to provide readers with the elegant designs of InterContinental hotels.

洲际酒店集团成立于 1777 年，有着 200 多年的历史，同时它也是目前世界酒店集团中最大的酒店管理集团，旗下拥有 8 大品牌酒店，在世界 100 多个国家拥有 4,480 家酒店，共 658,000 间客房，每年接待 15,300 万客人。庞大的数字真实的反映出洲际在全球酒店业的影响力。酒店的发展依靠的是全球化推动，从另一方面，洲际酒店的成长正是全球发展的一个缩影，而洲际的设计核心也正迎合了世界的发展需要。

当今世界，人们对可持续性环境的需要正在逐渐加大，洲际酒店针对这一需要意识到酒店面临的一个重要问题就是如何能在酒店设计中实现可持续化的生存发展，因此洲际正逐渐的转变酒店的设计重心。美国室内设计协会设计师、协会主管卡林·沃栽佑女士曾为洲际设计做出总结：酒店在设计中侧重环境意识，而在美感上做出了一些牺牲。洲际将环境意识带入酒店设计，要求设计师既考虑酒店对环境的影响，也要顾全酒店的优雅与舒适。在这两点产生冲突时，设计师应更多考虑它环境的影响，甚至是设计对未来环境的影响。洲际酒店不仅在观感上会产生悦目的效果，还是可持续性的，不会与生态环境产生冲突。在实施方面，酒店的建筑原材料提倡绿色环保，同时也会尽可能地减少建设能源消耗，从而保护环境，减少污染。

当今世界也是飞速发展，彰显个性的时代。如今传统的商务豪华型酒店，不再能满足商旅的要求。人们对时尚与艺术的追求传至各个领域，设计酒店的兴起为酒店设计注入新鲜的血液，并迅速引起人们的关注。而洲际也开始重视酒店设计的时尚感与艺术性。洲际酒店的设计从视觉及心理角度，会刻意营造出赏心悦目的感觉。特别在装饰造型上，创意新颖，独具一格。例如浦西洲际酒店，新加坡 LTW 室内设计公司为酒店的大堂设计了一个玻璃雕塑，特别的是一个雕塑作品被悬挂于天花板，仿佛一只飞舞的绸带，将刚到酒店的客人自然的引至接待处。此外，在洲际酒店，人们可寻新潮设计，从装饰色彩到家具造型，洲际具有超级的眼光，并引领潮流。

当今世界正经历着文化大融合的现象，经济贸易的往来与信息交流也得到发展，这让酒店成为一个传达、表现文化的聚集地。而洲际酒店无疑是传承文化的代表，酒店的设计将多元文化融合展现。例如人们会在充满现代设施的高端洲际酒店，见到中国景观园林的元素，甚至更深层次的园林意境，例如借景增加景深与层次，把自然元素引进室内，将室内与室外空间巧妙融合，创造出自然和谐的空间。

洲际酒店的发展显然是随着世界的脚步前进的，在洲际酒店看到的不是对建筑与室内设计亘古不变的标准与约束，而是不断改进、不断完善的设计要求。本章选取两家洲际旗下的品牌酒店，让视觉去体验洲际同世界一起成长的设计风采。

INTERCONTINENTAL PUXI

上海浦西洲际酒店

Situated in Puxi — the pulsating heart of Shanghai's cultural, commercial and residential centre — the InterContinental Hotel is the only five-star hotel in the vicinity. Like a beacon gleaming amidst the raw urbanity, this modern hotel offers tranquil respite from the bustling surroundings.

The hotel has 533 rooms with 25 Executive Suites, three Penthouse Suites, one InterContinental Suite and two Presidential Suites plus a spacious Club InterContinental Lounge. In addition, it features 3,500 square metres of flexible meeting space and a 350-seat multi-purpose auditorium; a 700-seat grand ballroom and 14 function rooms.

Majestic in scale yet boasting a clean and contemporary aesthetic, the hotel's interior is the ideal backdrop for the rich visual tapestry of design elements inspired by Chinese cultural symbols and handicrafts. With its 10-metre-high ceiling and grand structural columns embellished by intricate laser-cut patterns, the arrival lobby is a sight to behold. The columns resemble Chinese lanterns and create a rhythmic flow to the voluminous space. Floating overhead is a spectacular glass sculpture that is inspired by the Chinese ribbon dance. This bespoke art piece also serves as a

link to the reception lobby, which is anchored by a solid ebony wood counter, carved using traditional Chinese methods. Moving on to the lobby lounge, guests are made to feel welcome in this stately space that is made more intimate by cosy nooks and a lavish layering of patterns and details. Another stunning light feature to be found in the lobby lounge bar — a cascade of brass rods with crystal ball ends that form geometric Chinese motifs — takes centre stage here.

The design for the well-appointed guest rooms takes inspiration from the Chinese lotus, as seen in patterns used for the bed headboard. Refined elegance governs the colour palette, providing a soothing and restful environment for guests. With its open bathroom concept, the rooms feel remarkably spacious and airy. By separating the walk-in closet and toilet from the shower and bath areas, an easy sense intimacy is ensured.

A well-located hotel within the busy hub designed to offer not only all modern convenience and comfort to business travellers and tourists but also an environment where weary travellers can unwind and seek serenity.

Completion date: 2009
Location: Shanghai, China
Designer: LTW Designworks Pte Ltd

Photographer: Marc Gerritsen
Area: 10,375m²

完成时间：2009 年
项目地点：中国，上海
设计师：LTW 装饰设计有限公司

摄影师：汤马克
面积：10,375 平方米

浦西洲际酒店位于上海的文化、商业和住宅中心——浦西，是附近唯一的一家五星级酒店。酒店宛如粗糙的城市风格中一座闪闪发光的灯塔，为喧嚣的四周提供了宁静的世外桃源。

酒店拥有533间客房，其中包括25间行政套房、三间顶层套房、一间洲际套房和两间总统套房，还有一个宽敞的洲际酒廊。此外，酒店还配有3,500平方米的灵活会议空间、可容纳350人的多功能礼堂、可容纳700人的宴会厅和14间功能厅。

酒店规模宏伟，以简洁而现代的美学为特色。源于中国文化象征和手工艺的设计元素交织出丰富的视觉感。到达大厅值得一看，10米高的天花板和宏大的结构支柱采用交错的镭射切割图案装饰。形似中式灯笼的柱子营造出空间的韵律感。头顶悬浮的玻璃雕塑受到了中国缎带舞的启发。这件定制的艺术品同时还连接着前台大厅。大厅的乌木前台采用传统中式工艺雕刻，异常精美。大堂吧宏伟的空间被舒适的角落和丰富的图案和细节层次装点得更加贴心。大堂吧中央的另一个闪光点：末端为水晶球的黄铜条相互层叠，形成了中式几何图案。

设施齐全的客房的设计灵感来源于中国莲花，莲花图案被运用在卧室床头上。色彩搭配以精致的优雅感为主，为宾客提供了舒适而宁静的环境。开放式浴室概念让房间显得异常宽敞和轻快。隔开步入式衣橱和洗手间与洗浴空间被隔开，保证了轻松的亲密感。

这家位于繁华都市中的酒店不仅为商务人士和游客提供了现代舒适和便捷，而且让他们在此能够体验到放松和宁静的环境。

1. In the lobby, floating overhead is a spectacular glass sculpture that is inspired by the Chinese ribbon dance.
2. The stunning light feature to be found in the lobby lounge bar – a cascade of brass rods with crystal ball ends that form geometric Chinese motifs – takes centre stage here.

1. 大堂内，悬挂在头顶的华丽玻璃雕塑受到了中国缎带舞的启发
2. 大堂酒廊的灯具异常耀眼——末端为水晶球的黄铜条相互层叠，形成了中式几何图案

1. Wherever on the carpets or on the lightings, such as in the lift lobby, the traditional geometric Chinese motif appears.
2. The ballroom's interior is inspired by Chinese cultural elements, such as the wood screen.

1. 大堂电梯厅的地毯和灯具上都反复体现了中国传统图案元素
2. 宴会厅的设计从中国传统文化中获得了灵感，例如木屏风

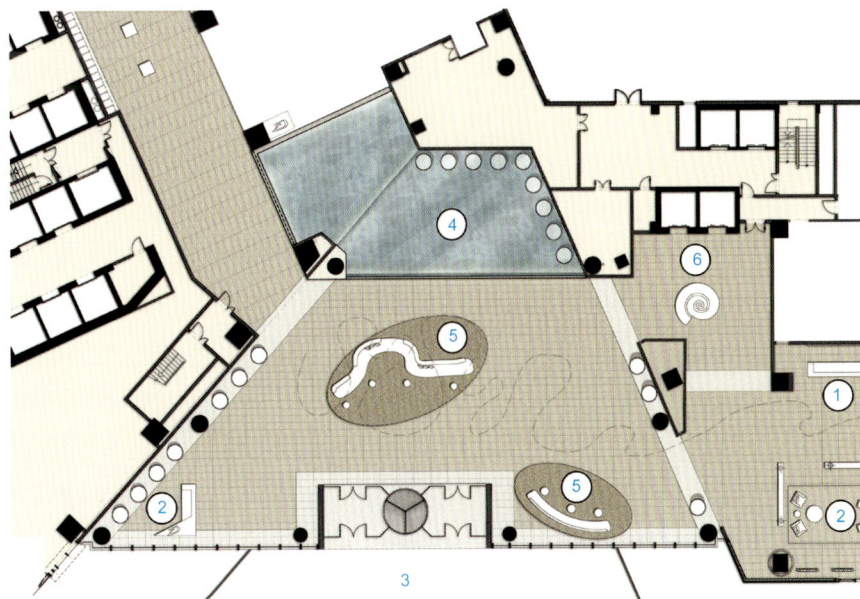

1. Reception
2. Concierge
3. Entrance
4. Reflection pond
5. Lounge
6. Lift lobby

1. 前台
2. 门房
3. 入口
4. 倒影池
5. 休息室
6. 电梯大厅

1. In Ecco All Day Dining Restaurant, separated booths, half concealed with lattice screens, provide ample space for more intimate.
2. The atmosphere in pre-function seating area blends luxurious with contemporary aesthetic.
3. Through the transparent wall and roof, the swimming pool resembles a crystal ball in the hotel.

1. 全日制餐厅的包房半隐藏在木格屏风后面，提供了足够的私密空间
2. 准备休息区在奢华中添加了现代感
3. 半透明墙壁和屋顶让游泳池宛如酒店内的一个水晶球

1

2

1. The study in presidential suite embellished with dark colour palette, matching with the modern art piece, shares a unique elegance.
2. The living room in presidential suite is equipped with the most advanced technology facilities, echoing with the modern atmosphere.
3. The elegant printing wallpaper refreshes the whole club room.
4. The presidential suite is in extremely simple style.
5. In the guestroom, refined elegance governs the colour palette, providing a soothing and restful environment for guests.

1. 总统套房的客厅以深色为主色调，搭配着现代艺术品，拥有独特的高雅风格
2. 总统套房的客厅配有最先进的技术设施，呼应了它的现代风格
3. 优雅的印花墙纸让整个俱乐部焕然一新
4. 总统套房采用了极其简单的风格
5. 客房采用了精致优雅的色调，为宾客营造了舒缓、宁静的环境

INTERCONTINENTAL REGENCY BAHRAIN

巴林洲际酒店

InterContinental Regency Bahrain called for a modern and elegant theme, with a distinctive touch of the Arabian. The design challenge facing dwp was how to respond to the considerable constraints presented by the existing structure, and overcome them to provide an elegant design, at once modern and respectful of the geographical and cultural location, as well as the history of the hotel. The original building structure proved, as time went on, to be in an even poorer condition, limiting the initial design proposals to ceilings and other structures, which would not be able to be executed. The other obstacle to overcome was the fact that the initial design concept was requested and accepted in 2006, but it took a few years before the hotel was ready to implement all of the changes. This meant that many selected fixtures, fittings and materials were no longer available on the market and further options have to be sourced, supplied and approved, to fit with the overall concept. Relishing the challenge, dwp seized the opportunity to establish new standards of sophistication in the Gulf, befitting such an exclusive hotel. The proposed modifications to the design concept were welcomed, as they maintained the original design intent.

The resulting design is fluid and adaptable, while maintaining an overall unique look for Bahrain. Much of the furniture and lighting was custom-designed, and the finest materials were sourced from around the globe, for timeless elegance. Rich regal colours and luxury materials and textures conspired to generate an ambience of refinement and superlative quality, with an opulent, yet restrained, Arabian flair. One of the key standout features, which marks the arrival for all visitors and guests to the hotel is a contemporary reception area with hallmark illuminated onyx reception desks, signifying a grand welcome. The creative environment was converted from a traditional and restricted space, into a modern and flowing, cleverly conceived arena, which conveys the brand identity and principles effectively.

A truly iconic hotel for the InterContinental Group in the region, the new interior design fits with the IHG design guidelines for their hotels, despite the existing building structure not being entirely compatible with their functional guidelines. The InterContinental Regency sets the benchmark for other hotels in Bahrain, having fully met client's expectations, while opening on time and within budget.

Completion date: March, 2011
Location: Manama, Bahrain

Designer: dwp | design worldwide partnership
Area: 12,200m²

完成时间：2011 年 3 月
项目地点：巴林，麦纳麦

设计师：dwp 设计世界合作事务所
面积：12,200 平方米

巴林洲际酒店以现代和优雅为主题，并且添加了独特的阿拉伯风情。dwp所面临的设计挑战是如何应对原有结构所呈现的大量限制，从而制作出优雅的设计。酒店既要拥有现代特色，又要尊重当地文化以及洲际酒店的历史。随着岁月的流逝，原有建筑结构日已破败，限制了初始设计中天花板和其他结构的执行。另一个障碍是初始设计概念早在2006年就获得了认可，但是这些年来一直没有得到实现。这意味着许多选定的设施、装饰和材料已经退出市场，设计师必须重新选择资源、供应商并获得认可。dwp抓住这次机会，在海湾地区这家独特的酒店建立了全新的精致设计标准。经过调整的设计概念大获欢迎，因为它们保留了初始设计的精髓。

最终的设计流畅而适应性强，同时又保留了巴林独一无二的整体外观。大多数家具和灯具都是特别定制的，而精致的材料则选自全国各地，经典而优雅。丰富的王室色彩与奢华的材质和纹理共同打造了精致而顶级的品质，豪华而节制，具有阿拉伯风情。当宾客走进酒店，现代的接待区和发光玛瑙前台将带给他们一个华丽的致意。整个空间从一个传统而局限的空间被改造成现代、流畅而构思精妙的舞台，有效表达了洲际酒店的品牌形象和经营理念。

巴林洲际酒店是洲际酒店集团在该区域的标志性酒店。尽管原有建筑结构与酒店的功能方向并不完全契合，全新的室内设计还是完美地呈现了洲际酒店集团的设计方针。巴林洲际酒店为巴林的其他酒店设计制定了基准，完全符合客户期望，并且兼顾了时间和预算。

1. In the night club, rich regal colours and luxury materials and textures conspired to generate an ambience of refinement and superlative quality, with an opulent, yet restrained, Arabian flair.
2. In the coffe shop, rich colours and luxury furnitures create elegant and delicate atmosphere.

1. 夜总会丰富的王室色彩和奢华的材质打造了精致而顶级的品质，同时又散发出富丽而节制的阿拉伯风情
2. 咖啡厅内，鲜亮的色彩，奢华的家具及其质地营造出精致文雅的氛围

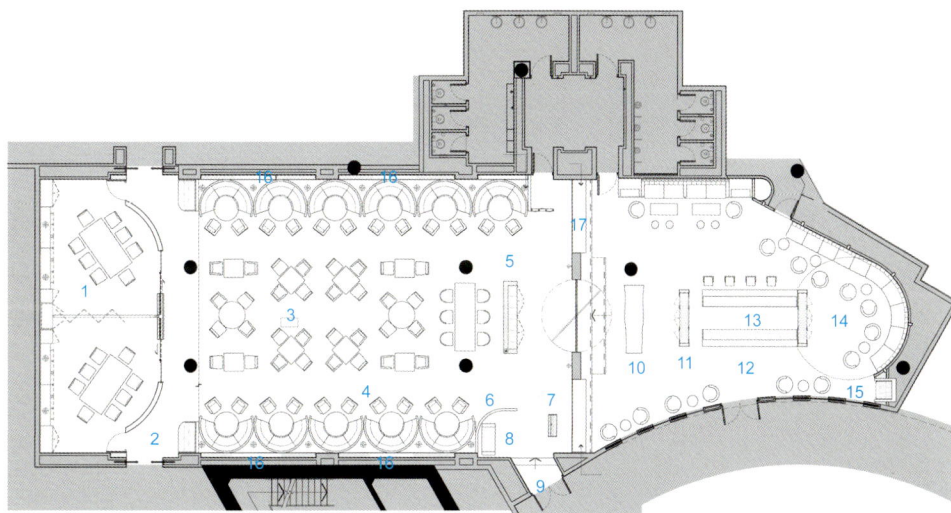

1. Private dining
2. Service & POS
3. Dining area
4. Booth seating
5. Display/Service counter
6. Screen

7. Maitre d' & POS
8. Waiting area
9. Entrance
10. 3,000mm long table
11. Wine display
12. Bar area

13. New bar
14. Wine display & feature above
15. Humidor
16. Back lit feature wall
17. White wine chiller

1. 包间
2. 服务室和POS机
3. 餐厅
4. 卡座
5. 展示/服务柜台
6. 屏风

7. 服务总管工作室和POS机
8. 等候区
9. 入口
10. 3米长桌
11. 酒架
12. 吧台

13. 新吧台
14. 酒架及上方特色展示
15. 保湿储藏室
16. 背光照明墙
17. 白葡萄酒冰酒器

1. The corridor next to the lounge is equipped with a long lattice as the wall, through which the lights exhibit a sense of hazy beauty.
2. With soft light shining on the silk blanket, the guestroom is bathed in elegant luxury.

1. 休息室旁边的走廊采用长木格墙隔断，灯光透过隔断形成了朦胧的美感
2. 柔和的光线照在丝绸毯子上，让客房尽显尊贵奢华

FOUR SEASONS HOTELS AND RESORTS
In the Name of Four Seasons, On Behalf of Guests

四季酒店集团——以四季之名，以人为本

FOUR SEASONS
Hotels and Resorts

Four Seasons Hotels and Resorts is a worldwide luxurious hotel company. Today, Four Seasons possesses 50 properties all over the world. The number is much less than that of other large hotel companies. Yet, Four Seasons is still regarded as one of the best hotels in the world, which is mainly relied on their excellent service and environment. The founder Isadore Sharp, who used to be an architect, attached architectural and interior design principles to hotel service. Four Seasons' design principle roots in their service philosophy "On Behalf of Guests". The guest-first hotel design ensures the real effects of great services.

In term of architectural design, with the wish to satisfy the guests' anticipation to "observe the customs of the place", Four Seasons pursues to combine hotel's architectural style to its surroundings, so that guests can experience the local culture and tradition within the hotel. In order to guarantee guests' safety, all the materials should be durable and every detail is well-constructed, without any hidden dangers. Four Seasons believes that a hotel with simple design yet well-selected materials and exquisite workmanship is responsible for guests, which avoids the potential dangers of architectural details. The layout of a Four Seasons hotel should be clear enough for guests to be aware of the organisation of functional areas. There should be minimum signage to guide guests in public area. Normally the public area is magnificent and intimate as well, which can easily adapt to the communication requirements of guests. In term of lighting design, as long as it is comfortable, both artificial lighting and natural lighting can be used. Four Seasons usually creates various atmosphere through the contrast between light and shadow in a focus area. Landscape areas and interior spaces will be decorated with artistic decorations to enhance the experience.

In interior design, guests' comfort is the priority. Highlighting the balance between interior design and architecture, Four Seasons tries to create distinctive design concept to provide guests with gorgeous and special lodging experience. The design styles are mainly Modernism, Minimalism, Classicism, Eclecticism or styles reflecting local culture. Four Seasons encourages the use of fundamental principles, such as the in-depth employment of proportion, geometry, volume, balance, contrast, colour, texture and detail. No matter which style and design principle is employed, the ultimate goal of Four Seasons is guests-oriented. The aim of extraordinary design is to exceed guests' expectation and please them. Four Seasons promotes meaningful design, not flubdubs. Four Seasons will add pleasant decoration or create a beautiful atmosphere in the aisles, ensuring guests to be pleased during their movements.

As its name suggests, Four Seasons cares about the seasons' changing and people's feelings and do everything "on behalf of guests". This chapter displays two hotels of Four Seasons to share their unique charms with readers.

四季酒店是总部设于加拿大的世界级豪华酒店集团。目前在全球拥有50家同品牌豪华酒店，这同其他大型酒店集团相比，在数量上相差甚远，但四季仍被认为是世界最佳酒店之一。这凭借的是酒店优质的服务与环境。酒店创始人 Isadore Sharp，曾是建筑专业毕业，他将建筑与室内设计的原则依附于酒店的服务原则。四季酒店的设计原则源自酒店的服务理念"以人为本"，以顾客为本的酒店设计可以将酒店的优质服务真正的落到实处。

在建筑设计上，从满足客人希望"入乡随俗"的期待，四季酒店追求将酒店的建筑风格与酒店所在地周围的建筑风格相融合，或从所在地的文化或建筑传统出发，让客人在酒店即能体会当地的民俗民风。为保证客人的安全，四季的建筑原料要求应是经久耐用，并且每处细节都应是施工良好，安全无隐患的。酒店认为一座设计简朴但是用料讲究，建筑工艺精湛的酒店是对客人安全最负责任的体现，减免了建筑细节产生安全隐患的风险。四季酒店的空间布局应是清晰的，以便顾客能轻松知晓酒店的功能区分配。并且应能以最少的方向指示牌来引导在公共区域接受服务的顾客。建筑内部公共空间一般应多给人以宏大并且亲切的感觉，以便适应顾客的社交需求。在照明方面，可以使用人造光或自然光，但以能提供舒适的照明为准则。四季酒店经常通过焦点区域充足的光线与较暗区域的对比，营造多样的氛围。此外，还会在景观区或室内设置艺术装饰，加强氛围的体验感。

在室内设计方面，顾客的舒适是最优先考虑的事情。四季酒店的室内设计在强调与建筑设计相协调的基础之上，会努力追求与众不同的设计理念，以为顾客打造赏心悦目且特别的住宿体验。设计的美学风格通常为现代主义，简约主义，古典主义，折衷主义，或者体现当地文化风格的设计。设计原则提倡运用基本原则，例如，在比例，几何形状，容量，平衡，对比，色彩，纹理及细节原则上，可以进行深入的运用。但是无论为那种美学风格和设计原则，四季的目标仍然是以顾客为准，与众不同的设计目的是超越顾客的期望并愉悦他们的感官，提倡有底蕴的设计，而不是哗众取宠。在细节上，四季酒店特别会在通道区域添置赏心悦目的装饰，或营造悦人耳目的氛围，让客人在移动的过程中体会愉快。

四季酒店，正如其名，如四季知冷知暖，以四季之名，展现以人为本的设计。本章展示了两家四季酒店，分享四季酒店设计的独特魅力。

FOUR SEASONS LOS ANGELES AT BEVERLY HILLS

洛杉矶比佛利山四季酒店

As a major hub for entertainment industry press junkets, celebrity events and charity fundraisers, the Four Seasons Hotel Los Angeles at Beverly Hills is a home away from home to an international list of celebrities, politicians and business executives. With the task of redesigning 285 guestrooms & suites, SFA's vision was to modernise the preexisting traditional-style interiors, while incorporating elements of classic Four-Seasons flora and invoking the rich history of Old Hollywood and Beverly Hills. With a warm, clean Southern-California style at its core, the design for Four Seasons Hotel Los Angeles at Beverly Hills evolved into one of classic aestheticism and smart simplicity.

The guestrooms' redesign nods to the timeless glamour of 1940s Hollywood, with elegantly mirrored wall panelling and white-gold contemporised chinoiserie headboards. The palette is composed of a soft blend of warm tans, fair woods and an elegant infusion of aqua. The tan carpet features a slightly abstracted aqua pattern of spherical flora, a modern tribute to the Four Season's long-standing tradition of incorporating the loveliest elements of nature. This same reinterpretation of floral-inspired patterns is apparent in smartly subtle details such as the petite, vibrant blue bolsters that bounce off the clean, white linens and the leaf-patterned throw at the foot of the bed. A matching throw pillow decorates the warmly upholstered, tan armchair with stitched in aqua details that simply hint at blooming foliage.

A unique scheme was developed for the One Bedroom Suite that was based on the overall design theme of Hollywood glamour updated to meet the needs of today's sophisticated traveller. A dramatic, fully upholstered, headboard wall in pearlised linen and a contemporised floral motif serves as the focal point above the bed. The woods are a rich warm, chocolate brown finish and the colour palette is a fresh mix of chocolate brown, ice blue and warm camel tones. The simple sophistication and transitional lines of the furnishings create a subtle cohesion between the suites and typical rooms.

In the Premier Suite, a colour palette of crisp spring greens, plums and warm silvery taupe tones is offset by rich chocolate wood finishes. The carpet was custom designed and inspired by the owners' love of flowers and lush gardens. While still attentive to the Hollywood glamour concept, the design for the Premier Suite focused highly on the needs of the modern guest, complete with a full dining area for entertaining and furnishings that are California fresh while incorporating transitional style.

Completion date: 2009
Location: Los Angeles, USA
Designer: SFA Design
Photographer: CSA Architects

Area: Renovation - 282 Guestrooms & Suites, Guest
Corridors, Lift Lobbies, Windows Bar & Lounge, Bar
Patio and Expansion, Ballroom & Pre-Function Space
and Library

完成时间：2009 年
项目地点：美国，洛杉矶
设计师：SFA 设计
摄影师：CSA 建筑事务所

规模：翻新——282 间客房和套房、客用走廊、
电梯大厅、窗口酒吧和休息室、露台酒吧和扩建、
宴会厅和准备室、图书室

作为娱乐业新闻发布会、名流活动以及慈善募集的中心，洛杉矶比佛利山四季酒店为来自全世界的名流、政治家和商界主管提供了家外之家。SFA对酒店的285间客房和套房进行重新设计，他们试图现代化先前存在的传统风格室内，同时融入经典的四季植物元素，唤醒旧好莱坞和比佛利山的丰富历史。设计以温暖、干净的南加州风格为核心，演变出经典的唯美主义和巧妙的简洁感。

客房设计体现了20世纪40年代的好莱坞经典魅力，设有优雅的镜面墙板和白金色现代中式风格床头。色彩搭配由柔和的暖棕色、精致的木色和优雅的水色组成。浅棕色地毯上带有抽象的水绿色球形植物图案，以现代方式呈现了四季酒店长期以来一直采用可爱自然元素。同样的植物图案也呈现在一些精致的细节上，例如白净床单上娇小而活泼的蓝色长枕和床脚抱枕上的树叶形图案。相匹配的抱枕装饰了温馨的棕色软垫扶手椅，上面的水绿色细节缝纫暗示着盛开的花朵。

单卧室套房的独特设计方案以好莱坞的闪亮魅力为基础，经过升级来符合现代旅行者的需求。夸张的软垫珠光布床头和现代化花朵图案是床铺的焦点。木材提供了丰富而温暖的巧克力棕色装饰。房间色彩以巧克力棕、冰蓝和温暖的驼色为主。简练精巧的设计与家具的过渡型线条在套房和标准客房之间营造了微妙的统一感。

高级套房采用了春绿色、梅红色和和银灰色为主色调，配以丰富的巧克力色木制装饰。特别定制的地毯体现了酒店对花朵和花园的热爱。高级套房的设计同样采用了好莱坞的闪亮概念，更加注重现代宾客的需求，配有娱乐就餐区，而装饰则兼具加州风情和过渡风格。

1. In the One Bedroom Suite, the furnishings are a rich warm, chocolate brown finish and the colour palette is a fresh mix of chocolate brown, ice blue and warm camel tones.
2. The carpet in the Premier Suite is custom designed and inspired by the owners' love of flowers and lush gardens.
3. The living room at Presidential East Suite features a modern design with a penthouse feel and walk-out balconies.

1. 单卧套房的装饰采用了丰富温暖的巧克力棕色，同时混合了冰蓝和暖驼色
2. 顶级套房的地毯经过了特别设计，从业主最爱的花朵和茂盛的花园中获得了灵感
3. 东总统套房设有一个阁楼式的可步入式阳台，使整体室内设计别具现代感

1. Reflecting Old Hollywood in an updated approach, the carpet in the Presidential West Suite is hand-tufted with an over-scaled, contemporary butterfly motif. Swarovski crystal sconces and chandeliers interpreted in a very modern way of leaves, branches and flora are suspended on cables and add a dramatic sparkle throughout the space.

2. The Royal Suite is a reflection of the beautiful rich tones and culture of the Orient. Wood flooring with inset carpet of bold contemporary floral motif in a chocolate background are the backdrop in the living area to create a luxurious residential feeling.

1. 西总统套房的地毯反映了旧好莱坞的风格，采用了特大的现代蝴蝶图案；施华洛世奇水晶灯座和吊灯由树叶、树枝和花朵组成，通过锁链悬挂，增添了生动的闪烁效果
2. 皇室套房汇集了丰富的色调和东方文化；木地板和带有大胆花朵图案的棕色地毯装饰着起居区，营造出奢华的居家感

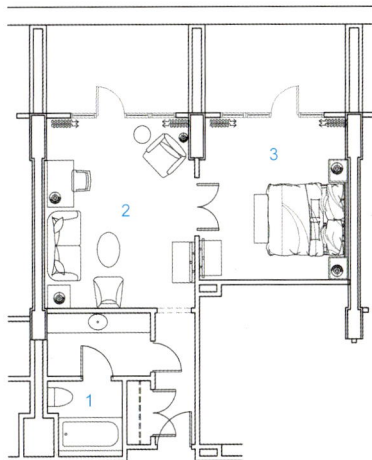

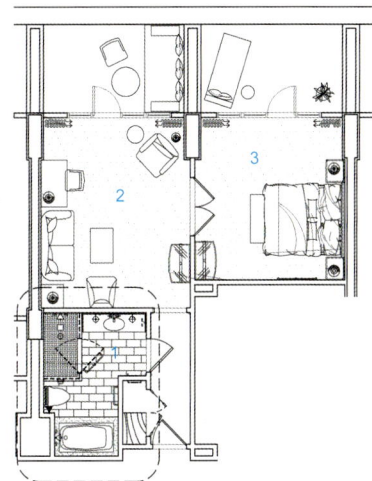

1. Bathroom
2. Living room
3. Bedroom

1. 浴室
2. 起居室
3. 卧室

1-2. The furnishings in Presidential West Suite reflect modern styling of Hollywood allure, with mirrored desks and coffee tables, cream and black lacquered nightstands and cabinets, and a graceful contemporary silver leaf dining table.

3. The colour palette of the Presidential East Suite is a contemporary mix of crème, chocolate brown and dusty blue in the living areas.

4. For the guest looking for a more modern design, the Presidential East Suite embodies clean lines, contemporary furnishings and bold contemporary artworks.

1、2. 西总统套房的装饰反映了好莱坞的现代风格，配有镜面书桌和咖啡桌、奶白色和黑色喷漆床头柜和橱柜以及优雅的现代银箔餐桌

3. 东总统套房的起居区混合了奶油色、巧克力棕和灰蓝色

4. 对追求更现代设计的客人来说，配有简洁的线条、现代的装饰和大胆的艺术品的东总统套房是最佳选择

1. The guestrooms' redesign nods to the timeless glamour of 1940s Hollywood, with elegantly mirrored wall panelling and white-gold contemporised chinoiserie headboards.

2. The colour palette of the Presidential West Suite is romantic, soft and elegant, imitating the soft tones of the butterflies, creams, plums, celadon greens and corals.

3. Another bedroom in the Presidential West Suite is outfitted with white-gold contemporised chinoiserie headboards.

4. The dramatic colour palette of Royal Suite creates for this suite includes blacks, chocolate browns, cinnabars and reds.

5. A dramatic, fully upholstered, headboard wall in pearlised linen and a contemporised floral motif serves as the focal point above the bed in Premier Suite.

1. 客房设计体现了20世纪40年代的好莱坞经典魅力，设有优雅的镜面墙板和白金色现代中式风格床头

2. 西总统套房的色彩搭配异常浪漫、柔和而优雅，汇合了蝴蝶色、奶油色、梅红色、灰绿色和珊瑚色

3. 西总统套房的另一间卧室采用了白金色中式床头

4. 皇室套房的夸张色调融合了黑色、巧克力棕、朱红色和红色

5. 装上软垫的床头墙采用了珠光亚麻布和现代花朵图案，是顶级套房的视觉焦点

FOUR SEASONS HOTEL DENVER

丹佛四季酒店

The Four Seasons Hotel Denver is part of a mixed-use 45-storey tower, housing the hotel as well as 102 private residences (floors 17-45) as "The Four Seasons Residences".

The hotel reflects a contemporary style, sympathetic to the environment with a strong sense of place. The materials selected are earthy and organic, both in colour and texture, a reflection of the rugged nature of Colorado. These elements are combined with a crisp, streamlined design concept appropriate for an urban hotel. The Guestrooms and Suites are welcoming as well as functional. There are three types of Standard Guestrooms, as well as eight One-bedroom Suites, two Two-bedroom Suites and one Presidential Suite.

The Public Spaces were designed on the premise that each area would flow to the next harmoniously. The artwork is a statement; it is bold in colour, content and scale. Most of these pieces were sourced from Colorado Artists and were selected to enhance the sleek, modern design concept. There are more than 1,000 pieces of original art through the public areas creating a gallery-like atmosphere.

The Bar & Restaurant have a sense of drama with the open plan layout and high ceilings, the intimacy of the fire-place and the use of rich woods and dark stacked stones add to this feeling. The wine display is the focal point and the unifying element which links the Bar to the Restaurant creating an immediate ambience from the moment of entry.

The hotel also has a Main Ballroom and Junior Ballroom, as well as a large Pre-function Area and several multi-purpose meeting rooms. These spaces have a unique identity with rich finishes in golds and purples and are enhanced with custom crystal light fixtures and specially designed carpets.

The Spa is an urban sanctuary in the heart of Downtown Denver. A palette of gold, cream and amber with aqua accents together with an interesting mix of unusual stones and mosaics and sustainable bamboo flooring creates an environment that is both calming and contemporary. The space is divided with a series of curved walls resulting in a smooth transition between intimate treatment rooms and male and female whirlpool areas.

Completion date: 2010
Location: Denver, USA
Designer: Bilkey Llinas Design (Interior Design Consultant),
HKS, Inc. (Executive Architect), Carney Architects (Design Architect)

Photographer: Riddle Don
Area: 25,060m²

完成时间：2010 年
项目地点：美国，丹佛
设计师：比尔凯·利纳斯设计（室内设计顾问），
HKS 公司(执行建筑师)，卡尔尼建筑事务所(设计建筑师)

摄影师：里德尔·唐
面积：25,060 平方米

丹佛四季酒店位于一座45层的商住混用大楼内。楼内除四季酒店外，还有102套私宅。

酒店反映了现代风格，拥有强烈的地方感和环境共鸣。酒店所选的材料在色彩和材质上都实在而根本，反映了科罗拉多州崎岖的自然景观。这些元素通过简洁流畅的设计理念融合在一起，适应了都市酒店的需求。

客房和套房热情四溢，并且具有功能价值。酒店有3种标准客房、8间单卧套房、2间双卧套房和一间总统套房。

公共区的设计以空间的流畅连接为前提。酒店装饰的艺术品在色彩、内容和规模上都十分大胆。它们来自于科罗拉多艺术家，用以提升时尚现代的设计主题。整个公共区域共有超过1200多件原创艺术品，营造了画廊一般的氛围。

酒吧和餐厅的开放式布局和高天花板极富戏剧感，而壁炉和丰富木材、黑色叠石的运用则进一步突出了这种感觉。酒架是设计的焦点，它连接着酒吧和餐厅，从一进门就营造出独特的即时感。

酒店拥有一个主宴会厅与一个普通宴会厅，同时配有宽敞的准备区和若干个多功能会议室。这些空间以金色和紫色的丰富装饰为特色，而定制的水晶灯具和地毯则进一步凸显了它们的独特身份。

水疗中心是丹佛市中心的一片世外桃源。金色、奶油色和琥珀色与水绿色相混合，与奇特的石头、马赛克和具有可持续性的竹地板共同营造出舒缓而现代的环境。水疗中心的空间被一系列弧形墙壁划分开来，在治疗室和男女漩涡池之间打造了流畅的过渡。

1. In any corner of the lobby, the artwork is a statement. There are more than 1,000 pieces of original art through the public areas creating a gallery-like atmosphere.

2. The artworks are bold in colour, content and scale, enhancing the sleek, modern, design concept of lobby.

1. 在大堂的任何角落，艺术品都十分显眼；整个公共区域共有超过1,000件艺术品，营造出画廊一样的氛围

2. 艺术品采用了大胆的色彩、内容和尺寸，提升了大堂的现代设计感

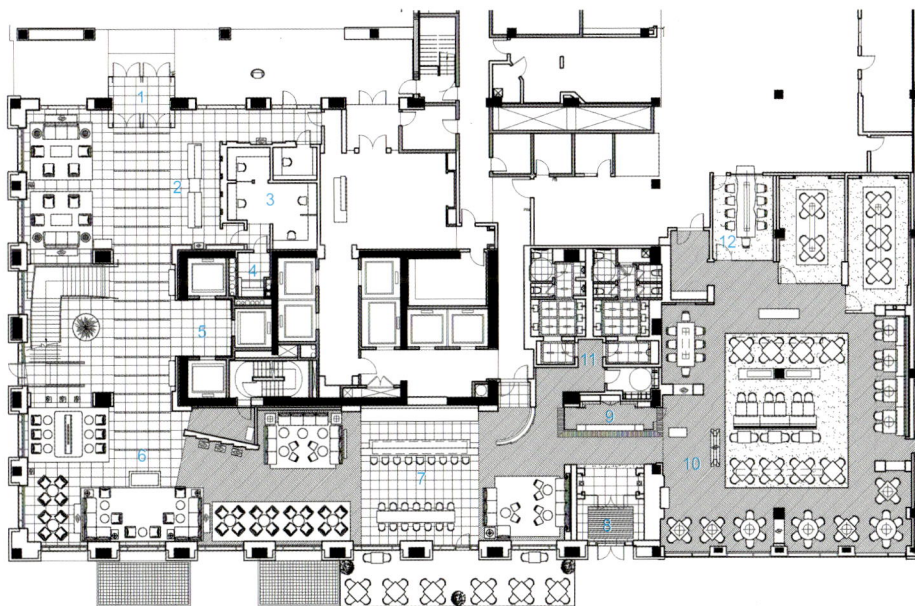

1. Main entry
2. Reception
3. Back office
4. Safe boxes
5. Lift lobby
6. Lobby lounge
7. Bar
8. Restaurant entry
9. Wine storage
10. Restaurant
11. Public toilets
12. Chef table

1. 主入口
2. 前台
3. 后勤办公
4. 保险箱
5. 电梯大厅
6. 大堂酒廊
7. 酒吧
8. 餐厅入口
9. 藏酒室
10. 餐厅
11. 公共洗手间
12. 主厨桌

1. The staircase in lobby is decorated with the piece from Colorado artists.
2. Dark palette and abstract paintings in Chef's table of EDGE enhance the dramatic feeling.
3. Grand Ballroom has a unique identity with rich finishes in golds and purples and are enhanced with custom crystal light fixtures and specially designed carpets.

1. 大堂楼梯装饰着由科罗拉多艺术家们设计的艺术品
2. 先锋餐厅深色的色调和抽象的画作提升了空间的戏剧效果
3. 大宴会厅拥有金色和紫色的丰富装饰，定制的水景吊灯和特别设计的地毯提升了空间效果

1. Spa entry, the entry to an urban sanctuary in the heart of Downtown Denver.
2. An interesting mix of unusual stones and mosaics and sustainable bamboo flooring creates an environment that is both calming and contemporary.
3. The spa salon continues the colour palette of the whole spa: gold, cream and amber with aqua accents.
4. Specially in Typical King Guestroom, the materials selected are earthy and organic, both in colour and texture, a reflection of the rugged nature of Colorado.

1. 水疗中心的入口，带你走进丹佛市中心的世外桃源
2. 各色稀有石头和马赛克有趣地结合在一起，与木地板共同营造出舒缓、现代的环境
3. 水疗沙龙延续了整个水疗中心的色彩：金色、奶油色和琥珀色
4. 标准国王房选择了质朴的有机材料，在色彩和材质上反映了科罗拉多州的自然特色

MARRIOTT INTERNATIONAL
Warm Design

万豪酒店集团——温暖设计

Marriott International is the largest hotel management group in the USA. From a beer shop founded in 1972, Marriott International has grown to be a leading lodging company with more than 3,600 properties with nearly 60,000 rooms in 71 countries worldwide. Marriott International still maintains a steady expansion tendency and stands as top three hotel brands for the last six years. This amazing success not only roots in its effective management strategy, but also in Marriott's focus on hotel branding culture. Furthermore, Marriott International makes consistent effort in hotel design. In the globalsation progress of nearly half a century, Marriot International remains grounded in a set of core values: put people first, pursue excellence, embrace change, act with integrity and serve our world. The "warm" design principle is the most impressive.

First of all, the warm design can satisfy the requirements from various guests, including the handicapped. Marriott tries its best to complete its barrier-free design and employs the latest design concepts in every newly built Marriott hotel, even the restored ones. Marriott International tries to complete its barrier-free facilities as much as possible and eliminates the existing barriers to provide warm services all round the hotels. Actually, the "barrier-free design" requirement of Marriott International is set before the hotel is built or purchased. In the early design phase, each new hotel must meet the local barrier-free regulations. Afterwards, Marriott International also adds some requirements, including: the hotel must provide a disabled passage for guests to access any area in the hotel; all the public toilets must provide toilets, urinals and commodes for the disabled; and in the hotel, at least 1% of the guestrooms need to be barrier-free guestrooms.

The warm design continues in the warm atmosphere of the interior design. Marriott International thinks it is mainly realised through design of public space and guestroom. In the design of public space, Marriott will emphasise some special areas to convey a warm experience visually and sensually. In the floor design, normally, on the base of combination of stone textures and carpet patterns, Marriott will choose stones with warm tones, instead of cold tones; the carpet patterns should be traditional and clear, with bright colours and warm tones. In the selection of wood works, wallpapers and window decorations in the public area, Marriott will select durable materials with high quality and appropriate scale. So the guest will feel at home although in a luxurious atmosphere. The furniture in the public area will also be qualified and durable, with a home style. The relevant fabrics are warm, fresh and bright coloured, avoiding any dim or weak colours. The colour palette tries to make a strong contrast. Besides, Marriott requires the hotel to use incandescent lighting in the public area to add warmth and comforts. In the guestroom design, there are some additional attention. First, the guestroom should be divided with several function areas — work, toilette, entertainment, snack and sleeping. Second, the art works in the guestroom should be close to the average appreciating levels of guests. For example, the themed realistic works can be easily accepted while the abstract and extra-modern art works should be avoided.

This chapter selects four hotels of Marriott International, including two Ritz-Carlton hotels and two JW Marriott hotels, displaying Marriott's high-quality warm design.

万豪酒店集团是目前美国最大的酒店管理集团。自 1972 年在华盛顿建立的第一家啤酒小店开始至今，万豪已在全球 71 个国家及地区拥有超过 3,600 家酒店，近 6 万个客房。万豪始终保持着稳定的酒店扩张趋势，连续 6 年位居全球酒店排名前三甲。这种令人惊异的态势不仅仅来源于有效的经营策略，除此之外，是万豪对酒店品牌文化的不懈追求与坚守。这其中之一，当然离不开万豪对酒店设计始终如一的苛求。万豪在近半个世纪的全球化历程中，一直对酒店的设计坚持着自己的品牌原则：温暖人心，引人入胜，有品位的奢华。这其中最令人感受至深的就是万豪的"温暖"设计原则。

首先，万豪的温暖设计原则体现在，设计能够满足各类客人的需要，包括残疾人。酒店集团努力的不断发展，完善酒店无障碍设计标准，将最先进的设计想法应用至整个万豪旗下的每个新建酒店，甚至每个翻新酒店，以至尽可能的改善酒店的无障碍设施，并排除现有障碍，将温暖的服务彻底贯穿于整个酒店。实际上，万豪的无障碍设计要求是在酒店建成或收购之前就预设规定的，每家新酒店的建立在设计之初，就必须符合当地国家所定的无障碍设计法规。之后在每个设计环节之中，万豪又有着一些在此方面的硬性要求，例如，酒店通道内需提供一个残疾人通道，可以允许有需要的客人能够方便进入酒店的任何区域；在酒店内所有的公共卫生设施需提供残疾人专用的座便器、小便器与洗脸台；酒店内，按比例至少 1% 的客房应为无障碍客房。

此外，这种"温暖人心"体现在对酒店室内设计的温暖氛围的营造上。万豪认为这主要是通过对公共空间以及客房设计而实现的。在公共空间的设计上，万豪会特别对这几个区域的设计向客人传达在"视觉及感觉"上的温暖体验。例如地板，万豪通常会在结合地板石材图案与地毯花色的基础之上，选择暖色石材，避免使用冷色；且地毯的花色应为传统并且清晰的图案，色调也应为明亮或暖色。再例如，在选择公共区域使用的木制品、墙纸或者窗饰时，万豪会选择经久耐用，高质量，比例适度的材料，以便使客人在感受到奢华的氛围之时，不产生对空间的距离感，而是感觉舒适亲和，就如回到自己的家中。公共区域的家具也要采用质量好、耐用，外观偏于居家形式的家具；相关的布艺也应为颜色温暖、清新、明亮，避免使用暗淡色或弱色，颜色间的搭配尽量产生对比效果。此外，有关公共区域的照明，万豪要求使用白炽光源以增加温暖，舒适的效果。在客房设计上，除保持在公共空间需要注意的以上几点，还应特别注意：第一，客房的空间设计应能分为工作、梳洗、娱乐、茶点、睡眠几大功能区域，几个区域需保持和谐、舒适、经久耐用的家居氛围。第二，客房内装饰的艺术品，应贴近客人的平均欣赏能力，例如有主题的写实艺术品通常易被接受，而抽象或过分现代风格的艺术品应避免。

本章选取 4 家万豪旗下酒店，其中包括两家 Ritz-Carlton 酒店和两家 JW Marriott 酒店，充分展示高品质的万豪温暖设计。

RITZ-CARLTON HONG KONG

香港丽思卡尔顿酒店

Towering over the vibrant cityscape at a staggering height of almost 490 metres, The Ritz-Carlton, Hong Kong, located in the 118-floor International Commerce Tower, proudly claims the title of the tallest hotel in the world. The iconic hotel dominates the topmost 16 floors and boasts stunning reception areas, an exquisite ballroom, six world-class dining venues and meeting spaces, as well as 312 luxuriously appointed guest rooms that offer state-of-the-art amenities and unrivalled panoramic views of the harbour and the city.

The Ritz-Carlton's signature refined elegance and intimate luxury is taken to new heights with an overarching design theme that reflects "Urban Living" and a strong sense of place. Hong Kong's dynamic contrasts — a bustling global metropolis at the crossroads of East and West, loaded with entrancing local flavour inspire the design language: an abstract yet seamless tale of modernity and tradition, set against a spectacular backdrop of stately opulence.

Entering the hotel on the 8th floor, guests get a first taste of the

Ritz-Carlton's legendary hospitality in the sprawling lobby lounge, which features a seating area and a cake boutique. A cool ocean of Palissandro blue marble stretches across the entire space here and throughout the hotel. In subtle contrast, the expansive wall is clad in warm-hued damask fabric panelled to create a fretwork screen that is reminiscent of Hong Kong's teeming urbanscape. Punctuating this pleasant rhythm is a dark "lacewood" timber entryway that leads guests through to the lift lobby that is counterbalanced with vivid orange onyx walls. At the end wall of each lift lobby is framed by important piece of artwork.

Throughout the hotel, modern interpretations of traditional Chinese fittings and art convey a distinctive style that is both familiar and inviting at the same time. To instill a sense of warmth and intimacy, and to demarcate areas in the larger public spaces, hand-tufted wool and silk "cartographic" carpets are used. With their jewelled blue-grey and earthy brown-yellow tones, the carpeted areas resemble floating parterres.

Heading up to the main reception lounge on the 102nd floor via a

Completion date: 2011
Location: Hong Kong, China

Designer: LTW Designworks Pte Ltd
Photographer: Marc Gerritsen, Jan Kudej

完成时间：2011 年
项目地点：中国，香港

设计师：LTW 装饰设计有限公司
摄影师：汤马克；简·库德吉

shuttle lift, guests will first encounter a soaring 8-metre atrium of gleaming marble. Overhead is an abstract depiction of a classic Chinese watercolour artwork of running horses. Fragmented strips of acrylic "dissect" the image and offer varying views of the artwork, inspiring awe and wonder from different angles.

With the help of strategic warm lighting, guests are naturally led to the more intimate check-in reception area, which is dominated by a sleek, hand-sculpted marble counter. Its organic motifs echo the lush carpet's cartographic theme. Behind the counter, the timber walls are padded with a faux antique-gold snake skin leather. The discreetly masculine feel of the space is softened by the use of a nature-inspired colour palette of warm autumnal gold and soothing tropical blues for the upholstery and carpet.

At the other end of the reception area, the Chocolate Boutique takes up an entire length of space with a stunning vista of the city. In this distinctly elegant yet cosy space, guests can partake in the Ritz-Carlton's signature afternoon tea. Comfortable furnishings flaunt classic clean lines and graceful curves, but with a disciplined modern edge. The main wall features a fretwork of antique bronze in three different finishes — its grid-like pattern is a subtle tribute to the sprawling urbanscape that the lounge overlooks.

With its scintillating "champagne and diamonds" theme, the ballroom — located on the 2nd floor — is a sight to behold. Radiating glamour, the 930-square-metre ballroom — one of the largest in the city — is graced by a dazzling sea of chandeliers with intricately woven crystals that seem to drip from the ceiling. Some of the wall panels feature cut-glass effects or an intricate fretwork pattern to enhance the ballroom's design concept. The colour palette here ranges from the ephemeral glow of champagne and amber to cool aquamarine and blushing fuchsia. To infuse this lavish space with chic comfort, some of the accent walls are clad in a lush cream fabric or textured leather; and outlined in dark marble. The grand lacewood doors that are fitted with cut crystal handles — these doors lead outside to a raised podium that is ideal for cocktail receptions. Bathed in natural light, the podium offers yet another breathtaking view of the city.

The hotel's design theme is carried through to the guest rooms and suites that occupy the 105th to 116th floors, all offering spectacular harbour or city views. Handsomely styled with tailor-made precision, the rooms offer refined comfort and the best modern amenities. A large Macassar ebony desk perches by the window, serving as an accent piece in the room. The plush armchair is outfitted in either champagne leather or velvet, depending on the room type. Behind the generous sized bed is a silk-panel wall and a leather headboard with metallic accents. Similarly, the TV wall is outfitted in leather and metal, featuring "baseball stitching" as a design element. The room is tastefully embellished with oriental accents — as seen in the floral motif of

香港丽思卡尔顿酒店盘踞在118层高的国际商务大厦楼上，以490米的高度俯瞰生机勃勃的城市环境，是全球最高的酒店。酒店占据着大楼最上面的16层，拥有绝妙的前台接待区、精致的宴会厅、6家世界顶级餐厅和会议空间以及312间奢华的客房。客房将为客人提供最先进的设施和海港与城市的无上美景。

在反映"城市生活"和强烈空间感的设计主题中，丽思卡尔顿酒店标志性的典雅和贴心的奢华中被提升到了新高度。香港的动态对比——一座位于东西方交界处的繁华的全球化都市承载着魅力十足的当地风情——为设计提供了灵感：在宏大而丰富的背景下，设计表达了现代与传统天衣无缝的结合。

从9楼进入，客人们会在大堂体验到丽思卡尔顿酒店传奇的热情好客。大堂以休息区和精品蛋糕店为特色。酒店的整个空间都被蓝金沙大理石的海洋所覆盖。广阔的墙面上包裹着温暖色调的装饰面料，形成了浮雕屏风，令人联想起香港热闹的街景。深色美国梧桐木入口通道引领着客人前往拥有橘色玛瑙墙面的电梯大厅。每间电梯大厅的端墙上都挂有一件重要的艺术品。

酒店内，传统中国装饰和艺术的现代解读表现出一种既熟悉又迷人的独特风格。为了增添温馨感和私密感，并且划分更大的公共区域，设计师采用了手工编织的羊毛和丝质"地图"地毯。宝石蓝灰色调和大地棕黄色调让地毯区看起来像飘浮的花坛。

通过电梯到达103层的主前台休息大厅，客人们将与高8米的大理石中庭见面。空中描绘出了一幅经典的中国画奔马图。分散的亚克力条对图像进行了仔细分析，呈现了艺术品各种令人惊叹的角度。

在策略性暖光的帮助下，客人被自然地引领到更私密的入住登记区。细长的手工雕刻大理石柜台上的有机图案与低碳的地图主题相互呼应。柜台后面的木板墙上包有人造仿古金色蛇纹皮革。而装饰物和地毯上温暖的秋天金色和舒缓的热带蓝色搭配则柔化了这种低调的阳刚气息。

在前台接待区的另一端，精品巧克力店占据了整面朝向城市风景的墙壁。在这个优雅而舒适的空间里，客人们能够享用丽嘉酒店标志性的下午茶。舒适的室内陈设以其经典简洁的线条和优雅曲线为特色，同时也兼具严格的现代特色。主墙上的古铜浮雕分为三块——它的网格状图案巧妙地呼应了休息大厅俯瞰的城市景象。

三楼的宴会厅以闪亮的"香槟钻石"为主题，值得一看。930平方米的宴会厅闪烁着光芒，是香港中最大的同类场所。水晶吊灯交织而成的炫目海洋似乎从天花板上滴落下来。一些墙壁板以雕花玻璃或复杂的浮雕图案为特色，提升了宴会厅的设计主题。这里的色彩从一闪即逝的香槟色和琥珀色延伸到了冷清的蓝宝石色和泛白的紫红色。为了给这个奢华的空间增添舒适感，一些重点墙被包上了奶油色织物或皮革，而外框则以黑色大理石包边。华丽的美国梧桐木门配有水晶把手——这些门通往露天平台，是举办鸡尾酒接待会的理想场所。沐浴在阳光之中，平台将提供另一种城市视野。

1. The check-in reception area is dominated by a sleek, hand-sculpted marble counter, of which organic motifs echo the lush carpet's cartographic theme.
2. The lobby lounge features a cake boutique, where guests get a first taste the Ritz-Carlton's legendary hospitality.

1. 入住登记前台最突出的就是光滑的手工打磨的大理石柜台，柜台上的有机造型图案与低碳的地图主题相互呼应
2. 大堂酒廊内有一家精品蛋糕店，客人可以率先品尝丽思卡尔顿酒店的传奇服务

the carpet and the bright tangerine silk-lined interior of the "Chinese jewel box" closet. The spacious bathrooms feature travertine marble floors and walls, and onyx countertops with double basins. The exclusive comfort that the Ritz-Carlton is recognised for is ensured in careful details — from the crystal-cut glass decanters to the unique artwork displayed in every room.

酒店设计主题被延续到了106到117楼之间的客房和套房之中，它们全部享有海湾或城市的优美景色。客房采用精准而时尚的设计，提供了贴心的舒适感和最现代的设施。根据房间的风格，长毛绒扶手椅采用香槟色皮革或天鹅绒包裹。宽敞的大床后方是丝质包层墙面和皮革金属装饰床头。同样的，电视墙也采用皮革和金属装饰，以"对缝法"为主要设计元素。房间有品位地添加了东方气息——体现在地毯的花朵图案和"中式珠宝盒"橱柜的亮橙色丝质内衬上。宽敞的浴室以石灰华大理石地面和墙壁以及玛瑙双水池台面为特色。丽思卡尔顿酒店独一无二的舒适感体现在精致的细节上——从水晶雕花玻璃瓶到每间房内独特的艺术品。

1. To instill a sense of warmth and intimacy, and to demarcate areas in the larger public spaces, hand-tufted wool and silk "cartographic" carpets are used.
2. With their jewelled blue-grey and earthy brown-yellow tones, the carpeted areas resemble floating parterres.

1. 为了植入温馨感和私密感，并且对公共空间进行划分，设计师采用了手工制作的羊毛和丝绸"地图"地毯
2. 宝石蓝灰和大地棕色系让地毯看起来像飘浮的花圃一样

1. Main ballroom
2. Function room
3. Prefunction area
4. Bridal room
5. Male restroom
6. Female restroom

1. 主宴会厅
2. 功能厅
3. 准备区
4. 新娘室
5. 男洗手间
6. 女洗手间

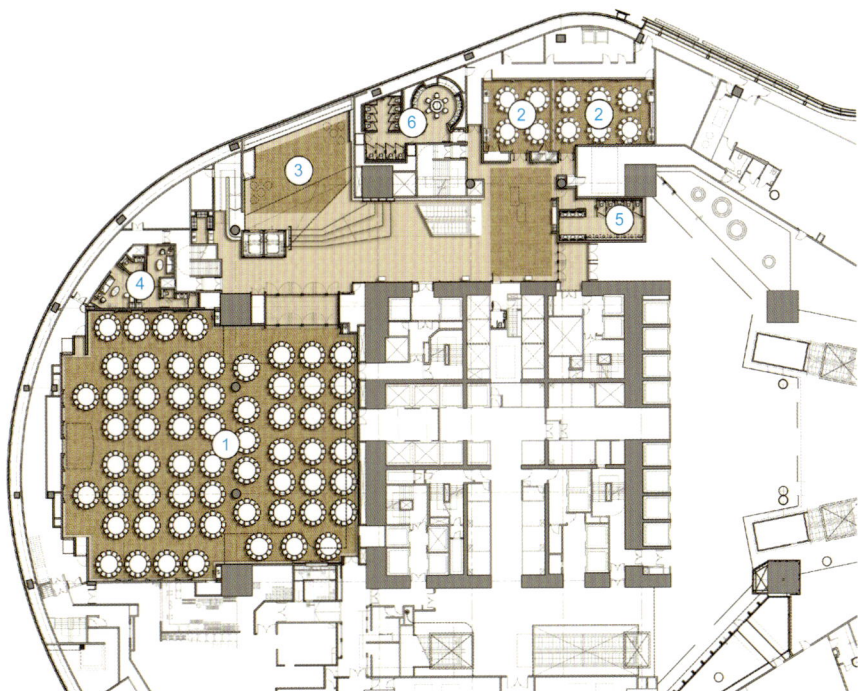

1. Radiating glamour, the 930-square-metre ballroom – one of the largest in the city – is graced by a dazzling sea of chandeliers with intricately woven crystals that seem to drip from the ceiling.
2. The lift lobby is counterbalanced with vivid orange onyx walls. At the end wall of each lift lobby is framed by important piece of artwork.
3. The lacewood timber entry way leads to the guestroom lift lobby. The piece of artwork and the carpet with cartographic theme creates strong artistic ambience.

1. 930平米的宴会厅是城中最大的宴会厅，闪烁在一片水晶吊灯交织而成的海洋之中，吊灯看起来像是天花板上滴落的水滴
2. 电梯大堂采用了鲜活的橙色玛瑙墙壁；电梯大堂的端墙上悬挂着珍贵的艺术品
3. 悬铃木入口通往客房电梯大堂；艺术品和地图主题的地毯营造出强烈的艺术氛围

2

3

1. The exclusive comfort that the Ritz-Carlton is recognised for is ensured in careful details – from the crystal-cut glass decanters to the unique artwork displayed in every room, even in the washroom.

2. The club lounge possesses an intimate, homey feel with the use of soft lighting, modern classic furniture, and shelves that display books that invite leisurely browsing.

1. 精心打造的细部设计保证了丽思卡尔顿酒店独有的舒适安逸——这些细节体现在水晶玻璃水瓶上，客房内独特的艺术品上，乃至洗手间里

2. 酒廊通过柔和的灯光、现代经典家居和书架打造了一种私密的家居感

1. With an ebony desk matching with the sophisticated lattice door, oriental luxury is created in Deluxe Suite.
2. The living room at Junior Suite is embellished with floral motif – the typical oriental accent.
3. The modern décor in the study of Presidential Suite employs a variety of lighting ways to enhance the luxury.

1. 黑檀木书桌与精致的格架门相互搭配，彰显了豪华套房的东方奢华韵味
2. 普通套房的客厅采用了花朵图案装饰，具有典型的东方气息
3. 总统套房书房的现代装饰利用不同的灯光来提升奢华感

1. The plush armchair is outfitted in velvet, and the TV wall features baseball stitching as a design element in the living room at Deluxe Suite.
2. The private dining at Presidential Suite is embellished with oriental accents – as seen in the bamboo lattice and the carpet with cloud motif.
3. The living room at Junior Suite is also embellished with oriental accents.

1. 豪华套房的客厅里，豪华的扶手椅配备了天鹅绒座套，而电视墙则以对缝缝纫元素为特色
2. 总统套房的私人就餐区采用了东方情调的装饰——正如竹子栅格和彩云图案的地毯所示
3. 普通套房的客厅同样采用了东方风情的装饰

1. Handsomely styled with tailor-made precision, bedroom at Deluxe Suite offers refined comfort.
2. Behind the generous sized bed is a silk-panel wall and a leather headboard with metallic accents.
3. Uniquely, the bathroom in Deluxe Suite is set in the centre of the suite surrounded by mirror walls and doors, which expand the space.
4. The spacious bathrooms feature travertine marble floors and walls, and onyx countertops with double basins.
5. The bathroom in Deluxe Suite is decked out with oriental lattice door.

1. 豪华套房的卧室经过了量身打造，精致而舒适
2. 特大的双人床后面是丝绸墙板，皮革床头上采用了金属装饰
3. 豪华套房的浴室被别出心裁地设在了正中央，四周环绕着镜面墙和房门，让空间显得更加宽阔
4. 宽敞的浴室以石灰华大理石地面和墙壁为特色，玛瑙台板上设有双重水池
5. 豪华套房的浴室采用了东方格子门

RITZ-CARLTON DUBAI IFC

迪拜丽思卡尔顿酒店

The partnership of HBA and The Ritz-Carlton has spanned 20 years and the completion of the DIFC location marks their 26th property. Located at the nexus of the financial district of Dubai, The Ritz-Carlton represents quiet luxury. The building covers an entire city block where HBA designed the 341-room hotel and 124 executive residences. The property also consists of 300 condominiums and luxury retail.

Inspired by grand European hotels like the Ritz in London and Paris, the 14-storey limestone building stands in strong contrast to the steel and glass skyscrapers that surround it. However, rather than looking back in time for inspiration, HBA moved forward to create an interior that is chic, streamlined and friendlier to the current generation of The Ritz-Carlton guests. Designers used the 1940s French Art Deco as the design style reference because the period represents chic luxury as well as more forward thinking in the design approach.

The colour palette is cool in contrast to the heat of Dubai. Things familiar to The Ritz-Carlton guests – fine crystal chandeliers, oriental carpets – are used but not necessarily in a traditional manner. Chandeliers in the Ballroom Wing and on the Mezzanine floor are inspired by jewellery designs from Cartier and Bulgari. The carpets in this area are very colourful designs of primarily Eastern inspiration, with border ornaments taken from jewellery designs.

The hotel main lobby is a tribute to the heritage of a renowned French Art Deco interior designer Jean-Michel Frank. Limestone walls and floors are accented with Frank-inspired straw pattern inlaid wood, while lift doors in polished metal feature his hallmark starburst pattern.

The first thing that meets the eye once entering the hotel lobby is a striking ceiling feature in azure Swarovski crystals by Lasvit. Not only does it add a distinct Art Deco feel to the hotel, but it also serves as a uniting element for the lobby, mezzanine and podium levels. Stunning in the daylight and at night alike, the "coral reef" still looks its best against the night sky.

There are three different guestroom plans with variations within each, 50 suites, including two Royal Suites, and a Club Lounge. The palette for the guestrooms and suites is pale ocean blue and gold with cream walls and Anegre wood detailing.

Completion date: January, 2011
Location: Dubai, United Arab Emirates

Designer: Sandra M. Cortner (HBA)
Photographer: Weng Ho

完成时间：2011 年 2 月
项目地点：阿联酋，迪拜

设计师：桑德拉·M·柯尔特纳
摄影师：翁和

The two Royal Suites are designed very much like 1940's Parisian Salons – the walls in the Parlour and Dining Rooms are upholstered in creamy silk in a pattern that matches the colour of the drapery. The study is wood panelled in Tiger wood and the bedroom has handmade area carpets and silk upholstered walls.

The Residences are operated by the hotel, but have a different feel from the guestrooms. All have open kitchens and a built-in look for a bespoke appearance and an efficient use of space. Even the palette for the service apartments differs from the guestrooms – warmer oranges and yellows with accents of celery green for a more residential feel.

设计师采用20世纪40年代法国装饰艺术作为风格参考，因为那个时代兼具时尚奢华和前卫的设计方式。

相对于迪拜的酷热来说，室内设计清爽干净。设计中体现了丽思卡尔顿酒店惯有的精致水晶吊灯、东方地毯等，但是以一种全新的方式呈现出来。宴会厅和中层楼的吊灯设计从卡地亚和宝格丽的珠宝设计中获得了灵感。这一区域五彩缤纷的地毯主要受到了东方的启发，而边缘装饰则参考了珠宝设计。

酒店大堂对知名法国装饰艺术设计师让-迈克尔·弗兰克表达了敬意。石灰石墙面和地面上点缀着弗兰克的镶嵌木稻草图案，而抛光金属电梯门则以他标志性的星光图案为特色。

进入酒店大堂，首先映入眼帘的就是以施华洛世奇水晶装饰的耀眼的天花板。它不仅为酒店增添了独特的装饰艺术感，还起到了连接大堂、中层楼和平台层的作用。水晶在白天和夜晚都会闪耀，这片"珊瑚礁"在夜空中格外美丽。

酒店拥有三种不同的客房、50间套房（包括两间皇家套房）和一个酒廊。客房和套房的色彩以淡海蓝色和金色为主，辅以奶油色墙面和红影木细部设计。

两间皇室套房的设计与20世纪40年代的巴黎沙龙十分类似——客厅和餐厅的墙面采用了奶油色丝质图案装饰，与帷帐的色彩相互搭配。书房的木护墙板采用了虎木，而卧室则拥有手工区域地毯和丝质衬垫墙面。

住宅同样由酒店运营，但是与客房的感觉截然不同。所有住宅都配有开放式厨房，并且对空间进行了有效利用。甚至连酒店式公寓色调都与客房大相径庭——它采用了更温暖的橙色和黄色，并点缀了芹菜绿色，更具家居感。

HBA与丽思卡尔顿酒店的合作已经持续了20年，而迪拜丽思卡尔顿酒店标志着他们的第26次合作。丽思卡尔顿酒店坐落在迪拜金融区的节点上，显得相当奢华。建筑占据了整个城市街区，HBA在其内设计了一家拥有341间客房的酒店和124套行政住宅。项目同时还拥有300套公寓和奢侈品店。

酒店受到了伦敦和巴黎丽思卡尔顿酒店等欧洲大酒店的启发，14层高的石灰石建筑与四周的钢铁玻璃摩天大楼形成了鲜明的对比。然而，HBA没有回溯历史寻找灵感，而是向前进打造了时尚、流线型并且更适合当代客人的室内设计。

1. The first thing that meets the eye once entering the hotel lobby is a striking ceiling feature in azure Swarovski crystals by Lasvit, which adds a distinct Art Deco feel.
2. The "coral reef" serves as a uniting element throughout the hotel, stunning in the daylight and at night alike.

1. 走进酒店大堂，首先映入眼帘的就是天花板上碧蓝的施华洛世奇水晶，整个空间充满了独特的装饰艺术氛围
2. "珊瑚礁"是贯穿酒店的元素，在白天和夜晚都炫彩夺目

1-2. In Centre Cut Restaurant, the focus is the border ornaments like a flock of birds flying to the air.
3. Old black and white photos hanged on the wall, No.5 Lounge and Bar brings guests to 1940s.

1、2. 中心餐厅的焦点是如小鸟飞向天空般的边界装饰
3. 墙壁上挂着古老的黑白相片，5号休闲酒廊将客人带回到20世纪40年代

1. The chic interior in No.5 Lounge and Bar
2. As seen in No.5 Lounge and Bar, designers used the 1940s French Art Deco as the design style reference.
3. The palette of the Spa is cool and the space serene, with white marble, glass mosaics and pale blues and greens.
4. The palette for the guestrooms and suites is pale ocean blue and gold with cream walls and Anegre wood detailing.

1. 5号休闲酒廊时髦的室内设计
2. 设计师在5号休闲酒廊的设计中参考了20世纪40年代法国装饰艺术风格
3. 水疗中心的色彩搭配十分清爽，采用了白色大理石、玻璃马赛克和浅蓝色与绿色装饰
4. 客房和套房以海蓝色和金色为主色调，搭配着奶油色墙壁和红影木细部装饰

JW MARRIOTT HOTEL BEIJING

北京万豪酒店

Located in the central district, the contemporary hotel anchors China's central place. Near the Olympic stadium and the Forbidden City, the 549-room 39-suite destination hotel conveys subtle sophistication and includes one of the largest ballrooms in the city, 1,800 square metres of adaptable meeting space. The hotel also offers an eclectic group of restaurants. Upon check-in at one of the five individual reception desks, guests can move into the rectangular shaped lobby lounge that has been transformed into a free flowing elliptical space with multiple seating areas and visually spectacular art installations. Within the guestrooms, high-tech amenities and soft ivory leather chairs are complemented by the pressed leather palates on the sliding doors. Vibrant contemporary paintings adorn the doors and pay homage to China's thriving art scene. Glass panels separate the bathroom from the bedroom, allowing natural light to radiate across the room and showcase the chic sunshine yellow and silver bathroom accessories. The open marble bathroom is completed with an LCD screen above the bath. For even more restorative benefits, the hotel includes an award-winning Quan Spa on the second floor, designed around the country's water culture. With dramatic raindrop sculptures coupled with chandeliers that float from the ceiling in a spiral of bubbles and iridescent azure curtains, shimmering like rain.

Rotating wine displays, a seafood bar and exposed meat aging room offers more opportunities for the interior design to enhance the dining experience. The open kitchen adds warmth to the interiors to contrast with the cooler and more minimalist lobby space adjacent to the restaurant. An undulating slat wall added to create more privacy as well as to minimise exposure to the daylight glare has enhanced the ambience of the space. The attached Loong Bar, with a dramatically lit dragon-shaped chandelier above the bar, is a WHAT spot for diners to enjoy before and after dinner drinks.

Completion date: 2008
Location: Beijing, China

Designer: HBA/Hirsch Bedner Associates
Photographer: HBA

完成时间：2008 年
项目地点：中国，北京

设计师：HBA/ 赫希贝德纳联合设计顾问公司
摄影师：HBA

1

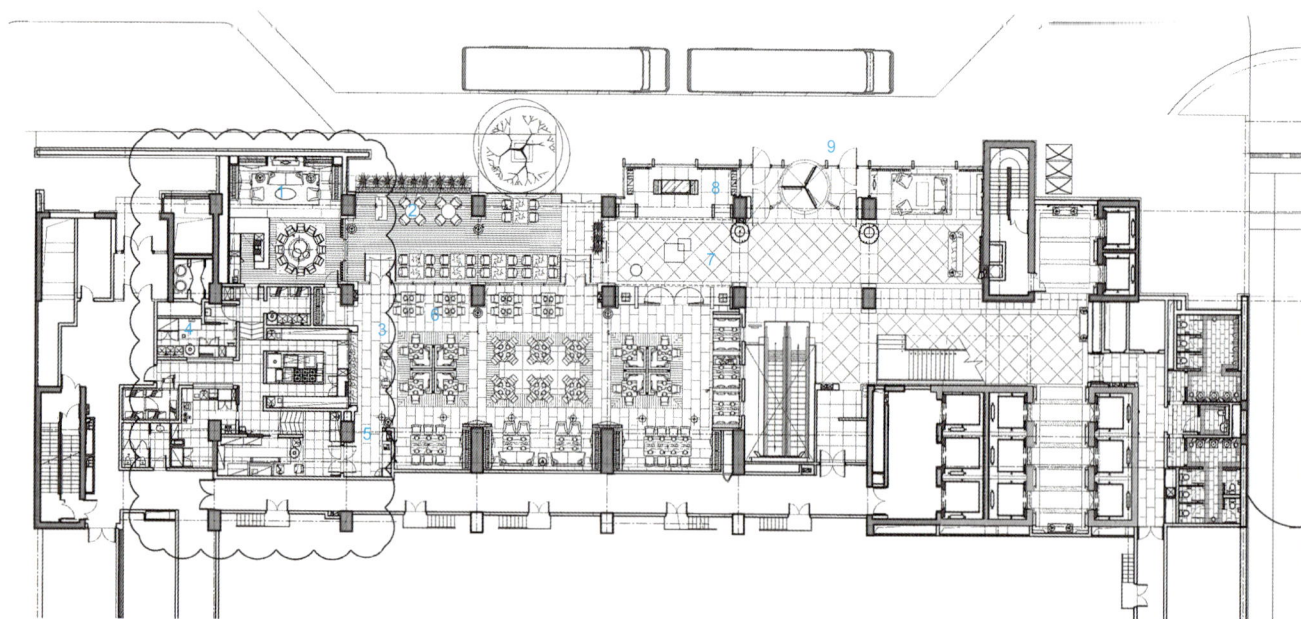

1. Private dining	6. Main dining	1. 包房	6. 主就餐区
2. Outdoor dining	7. Main entrance	2. 露天餐厅	7. 主入口
3. Service counter	8. Gift retail shop	3. 服务台	8. 礼品商店
4. Cold kitchen	9. Group entry	4. 冷厨厨房	9. 团体入口
5. Service station		5. 服务台	

1. Upon check-in at one of the five individual reception desks, guests can move into the rectangular shaped lobby lounge.
2. The private room in Pinot is embellished with Art Deco element.

1. 在五个独立的前台之一办理入住手续之后，客人将走进长方形大堂吧
2. 比诺餐厅的包房采用装饰艺术元素进行装饰

这家现代酒店位于中国中心（北京）的中心地区。万豪酒店紧邻奥林匹克体育场和紫禁城，拥有549间客房和39间套房，同时配有北京最大的会议场地，总面积达1,800平方米。酒店内设有多间不同风格的餐厅。在五个独立的前台之一办理入住手续之后，客人将走进长方形大堂吧。大堂吧现在已经被改造成为流畅的椭圆形，设有多个休息区并且摆放了大量艺术装饰。客房内的高科技设施和柔软的象牙白色皮革座椅与拉门上的皮革板相得益彰。活泼的当代画作装饰着房门，对中国欣欣向荣的艺术场景表达了敬意。浴室和卧室之间采用玻璃板隔开，让自然光能够穿过房间，突出了闪亮的金银两色浴室装饰。开放式大理石浴室的浴缸上方配有液晶屏幕。为了进一步提供健康服务，酒店三楼设有"泉"水疗中心，与中国的水文化息息相关。夸张的雨滴雕塑与枝形吊灯从天花板上成螺旋造型悬垂下来，配合着彩虹色的帷幔，如雨滴一般闪烁。

旋转酒架、海鲜吧和开放式肉类处理室为室内设计提供了更多提升餐饮体验的机会。开放式厨房为室内增添了一丝暖意，与简洁干净的大堂空间形成了对比。波浪起伏的板条墙提升了空间的私密感，减少了日晒，提升了整个空间的氛围。龙酒吧内的龙形吊灯栩栩如生，是餐前餐后喝一杯的理想场所。

1. Asia Bistro features elegant and relaxing ambience.
2. The private room at Asia Bistro
3. The open kitchen adds warmth to the interiors to contrast with the cooler and more minimalist lobby space.

1. 亚洲酒吧采用了优雅而放松的氛围
2. 亚洲酒吧的包间
3. 开放式厨房为室内增添了一丝暖意，与简洁干净的大堂空间形成了对比

1. CRU Steakhouse is specialised with its "flowing" feature walls.
2-3. The attached Loong Bar, with a dramatically lit dragon-shaped chandelier above the bar, is a WHAT spot for diners to enjoy before and after dinner drinks.

1. CRU牛排餐厅以其独特的"流动"墙壁为特色
2、3. 龙酒吧内的龙形吊灯栩栩如生,是餐前餐后喝一杯的理想场所

1. Noodles
2. Hot works
3. Display/Chef table
4. Salad/Cold cuts
5. Seafood
6. Roast duck
7. Bar/Beverages breakfast buffet
8. Semi private dining
9. Alfresco seating
10. Bakery retail
11. Retail shop
12. Bread dessert display
13. Fire shutters
14. Tandoor station
15. Dessert

1. 面条
2. 热菜
3. 展示/主厨桌
4. 沙拉/冷盘
5. 海鲜
6. 烤鸭
7. 吧台/饮料早餐自助
8. 半私密包房
9. 露天坐席
10. 面包房
11. 零售店
12. 面包甜点展示
13. 防火门
14. 印度泥炉烧烤台
15. 甜点

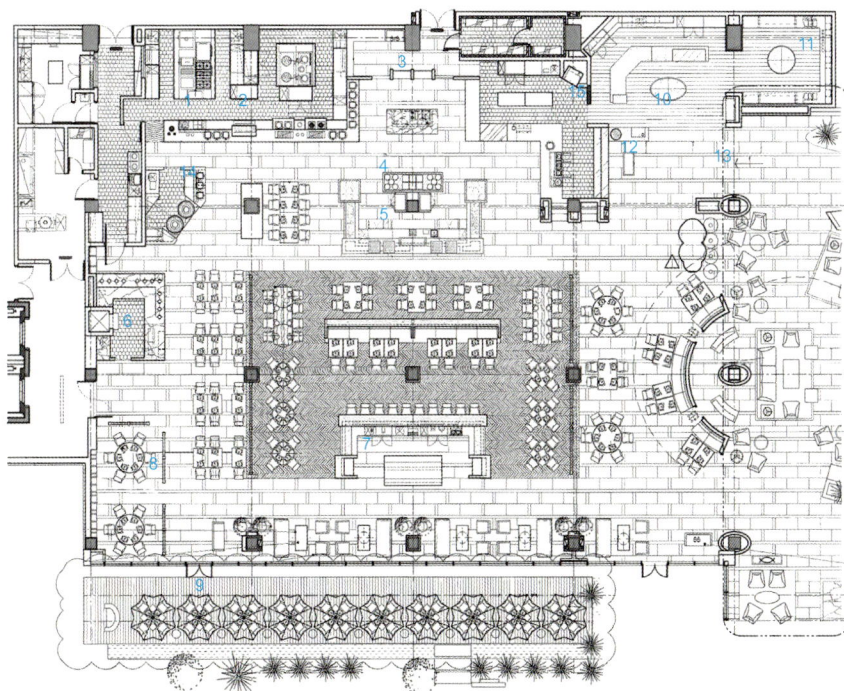

1. With its spacious space, Executive Lounge provides comfortable atmosphere.

2-3. The furnishings in Loong Bar are opulent, such as black sofa and blue chairs.

1. 行政酒廊宽敞的空间异常舒适

2、3. 龙酒吧内的装饰十分丰富，例如黑色的沙发和蓝色的座椅

1. With the colour palette of blue and yellow and the carpet with dragon motif, the interior in ballroom presents the royalty of Forbidden City.
2. The largest ballroom – Gala conveys subtle sophistication of the hotel.

1. 宴会厅以蓝色和黄色为主色调，配以巨龙图案的地毯，尽显紫禁城的奢华
2. 最大的宴会厅——节日厅体现了酒店的精致奢华

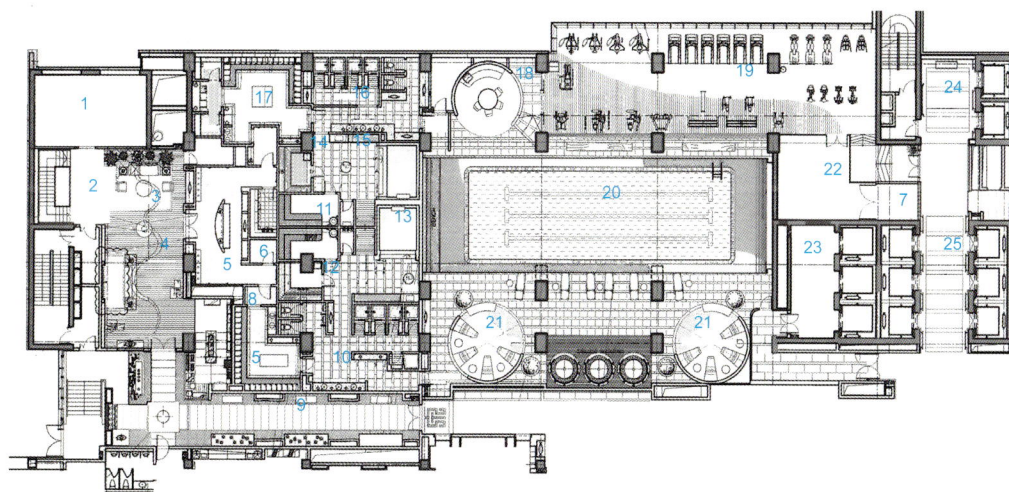

<table>
<tr><td>1. Hair salon</td><td>8. Toilet drop</td><td>15. Female wet area</td><td>22. Aerobics</td><td>1. 美发沙龙</td><td>8. 洗手间</td><td>15. 女浴区</td><td>22. 有氧运动区</td></tr>
<tr><td>2. Fire shutter</td><td>9. Public access corridor</td><td>16. Shower</td><td>23. Service lift</td><td>2. 防火门</td><td>9. 公共进入走廊</td><td>16. 淋浴区</td><td>23. 服务电梯</td></tr>
<tr><td>3. Waiting area</td><td>10. Dry vanity</td><td>17. Female locker room</td><td>24. Lifts to parking</td><td>3. 等候区</td><td>10. 化妆台</td><td>17. 女更衣室</td><td>24. 停车场电梯</td></tr>
<tr><td>4. Reception</td><td>11. Steam</td><td>18. Refreshment station</td><td>25. Lift lobby</td><td>4. 前台</td><td>11. 蒸汽区</td><td>18. 饮料站</td><td>25. 电梯大厅</td></tr>
<tr><td>5. Male locker room</td><td>12. Male wet area</td><td>19. Gymnasium</td><td></td><td>5. 男更衣室</td><td>12. 男浴区</td><td>19. 健身房</td><td></td></tr>
<tr><td>6. Steam plant room</td><td>13. Hot pool</td><td>20. Swimming pool</td><td></td><td>6. 蒸汽动力房</td><td>13. 热水池</td><td>20. 游泳池</td><td></td></tr>
<tr><td>7. Store</td><td>14. Sauna</td><td>21. Relaxation bench</td><td></td><td>7. 仓库</td><td>14. 桑拿区</td><td>21. 休息长椅</td><td></td></tr>
</table>

1. Within the guestrooms, high-tech amenities and soft ivory leather chairs are complemented by the pressed leather palates on the sliding doors.
2. Vibrant contemporary paintings adorn the walls and pay homage to China's thriving art scene.
3. The colour palette of the guestroom is mainly green and yellow, refreshing the tired guests.

1. 客房的高科技设施和柔软的象牙白色皮椅与压面皮革拉门遥相呼应
2. 充满活力的现代画作装饰着墙面，向中国欣欣向荣的艺术景象表达了敬意
3. 客房以绿色和黄色为主色调，让疲惫的旅客恢复了精神

JW MARRIOTT MARQUIS MIAMI

迈阿密万豪伯爵酒店

The client wanted this project to reflect the bold new brand and an upscale, five-star guest experience.

The concept was based on three key attributes: privileged, pampered and plugged-in. RTKL identified that guests are looking for these characteristics when choosing hotels. RTKL identified the "white space" in Miami's crowded hospitality market, creating new opportunities for an existing Marriott International brand. RTKL's demographic and psychographic analysis helped to define a targeted guest whose view of luxury travel included customised services, indulgent surroundings and high-tech amenities with a high-touch approach.

The hotel features neutral tones of the built-in, custom-designed wood desks, headboard, and other classic furnishings are accented by Axminster carpet woven in an oversized, bold pattern. Lighting effects feature soft down lights at the entry, dramatic wall sconces by the bathroom vanity, decorative nightstand lamps and built-in reading lights on the headboard. Bathrooms feature solid gold-coloured marble counters with carved-in sinks and polished Giallo Noce stone showers and baths. All adding to an overall feel of luxury.

The 41,806-square-metre JW Marriott Marquis Hotel offers 257 rooms and 56 suites with a design scheme that appeals to international business travellers and guests who prefer luxury travel. Public and social spaces are designed to encourage social networking and offer a sense of welcome. Other amenities include a two-floor entertainment complex, 929-square-metre NBA-approved basketball court, infinity pool, state-of-the-art fitness centre, virtual bowling alley, 1,858-square-metre ballroom, and the Jim McLean Golf School with putting greens and virtual golf simulators. In each room, guests will find additional touches of luxury including personal connectivity panels with multiple technology features, including the ability to connect personal technology devices to each room's massive LCD flat screen.

Completion date: October, 2010
Location: Miami, USA
Designer: Wendy Mendes

Photographer: Mike Butler
Area: 41,806m²

完成时间：2010 年 10 月
项目地点：美国，迈阿密
设计师：温迪·曼德斯

摄影师：麦克·巴特勒
面积：41,806 平方米

客户要求项目反映这个新品牌的理念，并且为客人打造高级的五星级酒店体验。

设计理念以三个主要特征为基础：特权、贴心和随时入住。RTKL认为，客人们在选择酒店时主要考虑这三个因素。RTKL发掘了迈阿密酒店市场的"空白领域"，为已存在的万豪品牌制造了全新机会。RTKL的人口统计和心理分析帮助酒店确定了客户群，他们寻求的奢华旅行中包括定制服务、舒适的环境和高科技个性化设施。

酒店以中性色调的嵌入式定制木桌、床头和其他经典家具为特色，配以带有大胆图案的艾克斯敏斯特地毯。在灯光效果方面：入口采用柔和的下射灯，浴室采用戏剧性的壁灯，而床头则采用装饰性床头灯和嵌入型阅读灯。浴室以金色大理石台面、刻入式水槽以及抛光石淋浴间和浴缸为主，尽显奢华。

这家总面积41,806平方米的万豪伯爵酒店拥有257间客房和56间套房。酒店的设计对偏爱奢华旅行的国际商务旅客和宾客极具吸引力。公共和社交空间的设计鼓励人们进行社交活动并给人以热情好客的感觉。其他设施包括一座两层楼的娱乐综合体、929平方米的美职联标准篮球场、无边界泳池、设备先进的健身中心、虚拟保龄球馆、1,858平方米的宴会厅和配有轻击区和高尔夫模拟器的吉姆·麦克林高尔夫学校。在每间客房内，客人都能享受额外的奢华，个性化连接面板具有多重技术特征，能够将私人技术设备与房间内的液晶屏幕连接起来。

1. Soft lighting shining on the neutral palette interior, ThreeFortyFive refreshes ordinary breakfast.
2. With several TVs throughout, Met Café & Bar is a typical example to prove that the hotel is designed to encourage social networking.

1. 柔和的光线与中性的色彩相搭配，3-40-5餐厅颠覆了传统的早餐理念
2. 遇见咖啡吧内设有若干台电视，证明了酒店促进社交网络交流的目的

2

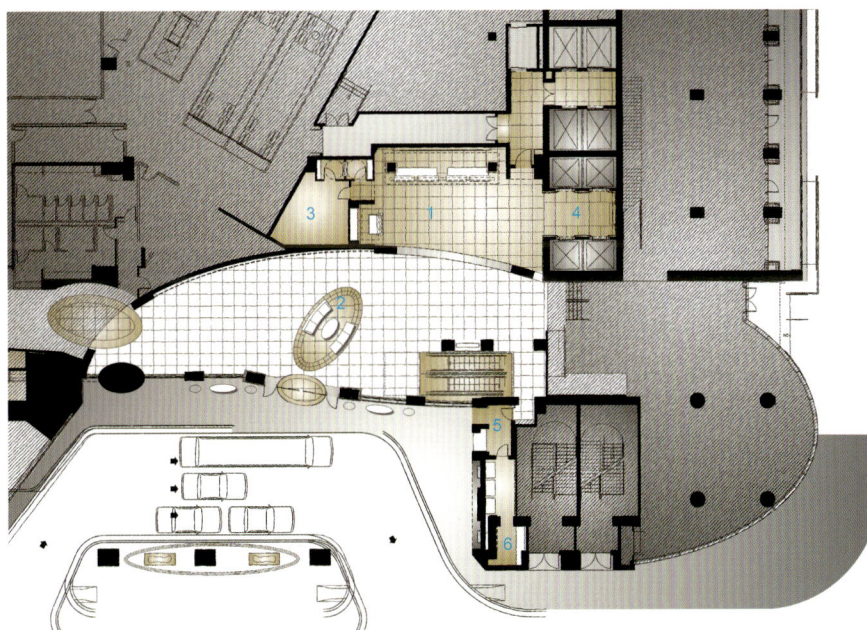

1. Reception
2. Lobby
3. Offices
4. Lift lobby
5. Valet
6. Luggage storage

1. 前台
2. 大堂
3. 办公室
4. 电梯大厅
5. 服务台
6. 行李寄存处

1. Featuring exquisite interior design of details, db bistro Moderne creates unique luxurious experience.

2-3. The living room in the guestrooms features custom-designed wood desks. And other classic furnishings are accented by Axminster carpet woven in an oversized, bold pattern.

1. 细节精致的室内设计让现代db酒吧营造出独有的奢华体验

2、3. 客房客厅内摆放着一张木制书桌。另一件经典装饰是图案大胆的艾克斯敏斯特地毯。

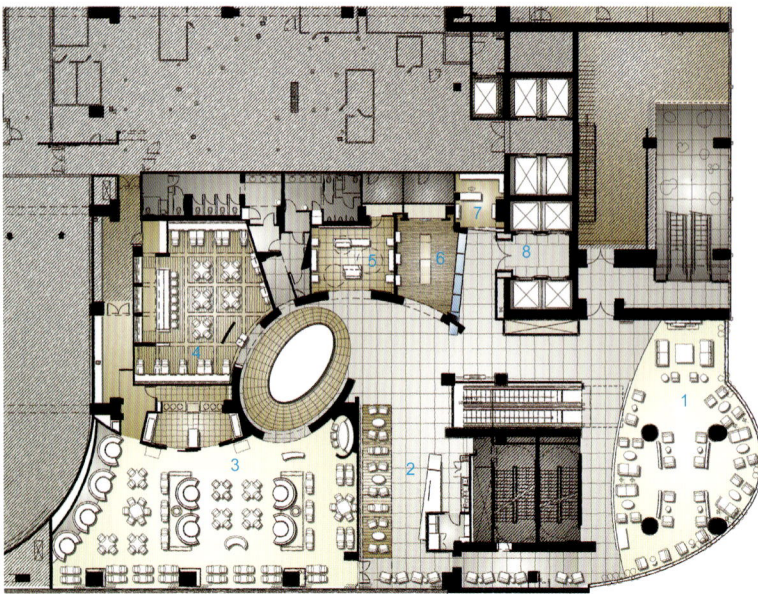

1. Vue Lounge
2. Intermezzo Café
3. ThreeFortyFive
4. Met Café & Bar
5. The Essentials
6. W Wine Boutique
7. Le Chocolatier
8. Lift Lobby

1. 休闲酒吧
2. 间奏咖啡厅
3. 3-40-5餐厅
4. 遇见咖啡&酒吧
5. 要素休息室
6. W精品酒廊
7. 巧克力商店
8. 电梯大厅

1. Opulent artworks add a sense of art.
2. A comfortable corner in the guestroom
3. Lying on the king-size bed and overlooking the view of the modern city through floor-to-ceiling windows, the guestroom offers additional touches of luxury.

1. 丰富的艺术品增添了艺术感
2. 客房中舒适的一角
3. 躺在特大双人床、透过落地窗远眺现代城市风景，客房显得极致奢华

HILTON WORLDWIDE
The Smiley Pioneer

希尔顿酒店集团——微笑的开拓者

Hilton Worldwide is one of the most well-known hotel groups in the world. Today, Hilton Worldwide possesses 540 hotels and 193,000 guestrooms all over the world. The steady development is not only reflected in the increasing numbers, but also the development of the hotel's brand culture. Now Hilton Worldwide is regarded as the most stylish and innovative hotel management group.

At the beginning of the first Hilton hotel, Conrad Hilton experienced in the hotel as a guest to feel about everything in the hotel. He found that "smile service" is an essential secret weapon in hotel development. This kind of service has gradually become a brand culture of Hilton hotels and is carried out in every staff's mind and behaviours, building a unique image of "smile". All the staff of Hilton have been taught that they should use "smile service" to make guests feel at home, including listening and accepting guests' advices with smile and exploring the insufficiency in every aspect with smile, which helps to complete the functions of the hotel architectural system. In this aspect, Hilton Worldwide is always leading the other hotel groups.

First of all, Hilton built its first comprehensive hotel through this smile culture. According to guests' requirement, the hotel created a complete architectural system. Besides providing spaces of board and lodging, Hilton took the lead in setting facilities like cafés, meeting rooms, ballrooms, swimming pools and a shopping centre in the hotel, and even introduced some other service facilities, including post office, flower shop, clothes shop, agency of airlines, travel agency and taxi stand. Next, Hilton clarified the guestrooms, subdividing them into single room, double room, suite and deluxe suite, which made Hilton the first hotel to provide deluxe suite for heads of states. The restaurants are subdivided into refined restaurants and fast food restaurants. In addition to the increase of functional areas, Hilton hotel learned from their smiley enquiry that the interior facilities also need to be completed. Therefore, Hilton became the first hotel to install air-conditioners in all the rooms. Besides, the rooms are also introduced with wine cabinets, telephones, colour TVs, radios and refrigerators. Since then, Hilton created their signature "feel at home" image through "smile service". Hilton continued to develop with their smiles in the almost complete architectural system. In 2004, Hilton created a unique leisure atmosphere in the interior design. The hotel allotted a lounge for guests to be refreshed, which is the first in hotel industry. The lounge is decorated with various lighting facilities, which can be adjusted according to requirements; flowers and fresh fruit will create a relaxing atmosphere both in the sense of vision and smell.

Today, the smiley pioneer – Hilton Worldwide is making continuous progress in completing hotel design. This chapter selects four Hilton hotels to explain the Hilton's successful models in detail.

希尔顿酒店集团是目前世界范围内知名度最高的酒店集团之一。目前集团已经拥有540家酒店，将近193,000间客房，产业遍布全球。稳步的发展趋势不仅体现在酒店数量上的增长，更体现在酒店品牌文化的发展上。由于对酒店建筑与设计尽善尽美的追求，希尔顿集团如今被评为最具风格与创新精神的酒店管理集团。

在希尔顿酒店创建之初，希尔顿曾以顾客的身份亲身在酒店体验，感受酒店的一切。他发现"微笑服务"是酒店发展必不可少的秘密武器。这种服务慢慢成为后来希尔顿酒店的品牌文化。这种品牌文化贯彻到每一个员工的思想和行为之中，从而塑造了独特的"微笑"品牌形象。希尔顿饭店的每一位员工都被谆谆告诫：要用"微笑服务"为客人创造"宾至如归"的文化氛围、其中包括以微笑听取或接纳顾客的意见，以不断的微笑去探寻酒店方方面面的不足，而这正起到了完善酒店建筑系统的作用，在这方面，希尔顿酒店通常引领其他酒店集团。

首先，通过这种微笑文化，希尔顿建成第一个综合性酒店。酒店根据客人的需求，在酒店内尽力创作一个完整的建筑系统。除了提供食宿的空间之外，希尔顿酒店领先其他酒店在酒店内设置咖啡室、会议室、宴会厅、游泳池、购物中心等设施，甚至还设置其他服务机构，包括邮电、花店、服装店、航空公司代理处、旅行社、出租汽车站等。接下来，酒店将客房设置更为细化，分为单人房、双人房、套房，更成为第一家为国家首脑级官员提供豪华套房的酒店。餐厅也被细化为高级餐厅和快餐厅。除了建筑内功能区的增加，希尔顿酒店还通过不断的微笑问询与虚心接受了解到，室内设备同样需要完善，希尔顿又成为了第一家在所有房间设置空调的酒店。除此之外，酒柜、电话、彩色电视机、收音机、电冰箱这类设备也被第一次纳入酒店的室内空间。从此，希尔顿酒店用他们的"微笑文化"创立了独特的"宾至如归"招牌。随后，希尔顿在基本完善的酒店建筑系统之中，仍不断的以微笑开拓着。在2004年，希尔顿在酒店的室内营造出独特的休闲氛围。酒店单独设计出为顾客提供恢复体力与精神的休息室，首开休闲之风。休息室内装饰多样的照明设备，并可调节，或明或暗；鲜花搭配着新鲜水果，无论从视觉和嗅觉都营造出放松的气氛。

如今，微笑的开拓者——希尔顿酒店，仍在完善酒店设计的道路上不断前进。本章选取4家希尔顿酒店为您详解希尔顿酒店的成功范例。

THE SKIRVIN HILTON OKLAHOMA CITY
希尔顿斯科文俄克拉何马城酒店

The historical Skirvin Hilton Oklahoma City is located at the heart of city centre and entertainment district of Oklahoma City. It underwent a multi-million US dollar renovation to restore the original grandeur that made the Skirvin Hotel an Oklahoma City landmark for the last 100 years.

It has 225 exquisitely appointed rooms, 20 Rotunda suites, a presidential suite, 1,720 square metres of meeting and banquet space, and a 570-square-metre grand ballroom.

All the design inside the hotel went through numerous coordination and certification process with the historical architecture specialists. Every corner and every detail was carefully studied before the designers combined the past and the present, with the original elegance remained and its classy atmosphere was showed through modern techniques.

For over 100 years of operation, the Skirvin Hotel has been synonymous with elegance and innovation, hosting oil barons, dignitaries, political leaders, and presidents, which made it a signature building standing proudly in the centre of Oklahoma City. What it represents is a status of nobility and fame, a status that is irreplaceable.

The 225 exquisitely appointed rooms at the Skirvin Hilton Oklahoma City have been designed with all of the comforts and conveniences of modern hospitality.

Completion date: 2007
Location: Oklahoma, USA

Designer: Turner Duncan, Kimberley Miller
Photographer: DFW Photobooths

完成时间：2007 年
项目地点：美国，俄克拉何马市

设计师：特纳·邓肯；金伯利·米勒
摄影师：DFW 摄影工作室

这座具有悠久历史的希尔顿斯科文酒店位于俄克拉何马市市中心的娱乐区。它经历了数百万美元的改造，对作为俄克拉何马市百年以来的地标的斯科文酒店进行了修复。

酒店拥有225间精致的客房、20间圆厅套房、一间总统套房、1,720平方米的会议和宴会空间以及570平方米的大宴会厅。

酒店内的所有设计都经过与建筑历史学家的无数次协商和认证。在结合过去和现在之前，每个角落、每个细节都经过了反复研究，保留了酒店原有的优雅并将它经典的氛围以现代技术展示出来。

经过100年的运营，斯科文酒店以其优雅和创新风格招待了石油大亨、高官、政治领袖和总统，使其成为了俄克拉何马市中心一座闪耀的建筑。它代表着难以替代的尊贵和荣耀。

希尔顿斯科文酒店的225间精致客房将为宾客们提供极致的舒适和现代酒店的便利。

1. Employing dramatic colour palette of red and blue, the lobby offers grandeur ambience.
2. Park Avenue Grill Lounge is designed in rich Art Deco.

1. 大堂采用了夸张的红色和蓝色，营造出宏大的氛围
2. 派克大街餐厅采用了丰富的装饰艺术设计

1. The interior design of Park Avenue Grill keeps original southwestern décor.
2. With the elegant arch door welcoming distinguished guests, the latest renovation keeps the original grandeur.

1. 派克大街餐厅的室内设计保持了原始的西南部风格装饰
2. 优雅的拱门引领着贵宾，翻修工作保持了原有的宏伟风格

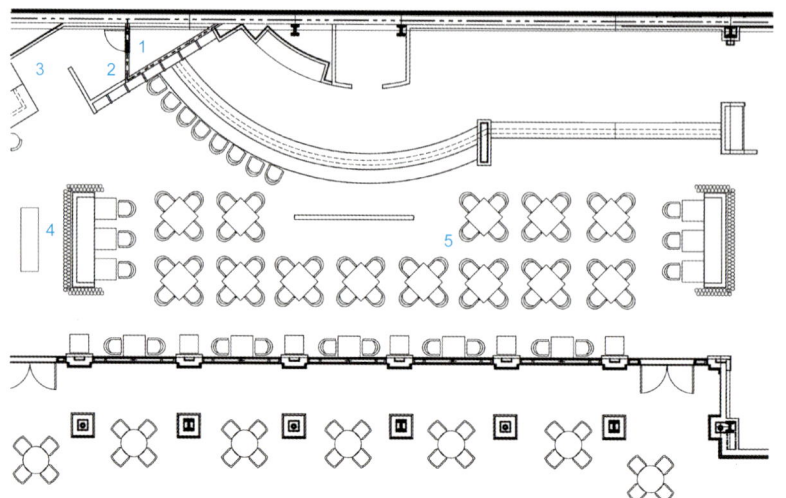

1. Chair closet
2. Service station
3. Service pick-up
4. Drink table
5. Restaurant

1. 座椅壁橱
2. 服务台
3. 餐车
4. 酒桌
5. 餐厅

1. The grandeur ballroom promenade
2. The ballroom foyer combines the spirit of design both from past and present.
3. The Grand Ballroom features sophisticated chandeliers, indicating historic luxury.

1. 华丽的宴会厅走道
2. 宴会厅门厅结合了过去和现代的设计元素
3. 大宴会厅以精致的吊灯为特色，展示了历史的奢华

1

1. Founders Room offers classy atmosphere with classic furnishings and antique artworks.
2. The colour palette of interior in guestroom offers noble atmosphere.
3. The Presidential Suite bedroom shares elegant and lively palette and luxurious textured fabrics on the bed.
4. Billiards Room is set with European style billiard table.

1. 创始人房以古典装饰和古董艺术品打造了古典的氛围
2. 客房的色彩搭配营造出尊贵的氛围
3. 总统套房拥有优雅且充满活力的装饰色彩以及奢华的印花床饰
4. 台球室设有欧式风格台球桌

HILTON GUANGZHOU TIANHE

广州天河新天希尔顿酒店

Conveniently located in Guangzhou's central business district, Hilton Guangzhou Tianhe is just 40 minutes drive from Baiyun International Airport and a few minutes walk from a transport hub that provides a metro ride to the airport, speedy train and bus links to Shenzhen and Hong Kong. The central location allows easy access to the city's most popular shopping and entertainment venues. Guests can enjoy the best of what Guangzhou has to offer while never straying far from the comforts of their hotel.

The 504 spacious guest rooms and suites offer the very latest amenities and comforts designed to make your stay carefree and relaxing. Each room is tastefully decorated in contemporary design, while floor-to-ceiling windows provide unrivalled views of the city's skyline. Wireless internet access, high definition TV and DVD player round out the Hilton vision of giving guests innovative retreats that embrace a sense of well-being.

Six restaurants and bars provide various delicacies. Sui Yuan with 13 private dining suites pays homage to the region's finest Cantonese cuisine. Il Ponte' presents authentic Italian cuisine in a modern and chic setting, with two private rooms. Café@2 is an all-day dining hub offering highly specialised breakfast, lunch and dinner buffets. Bar One is a relaxing retreat for meeting friends or associates, with live music in the evenings. T Lounge & Bar in the lobby serves refined afternoon tea and cakes amid the social whirl. Alfresco serves tapas, local snacks and refreshing drinks in a convivial atmosphere.

With 1,900m^2 of flexible meeting and banquet space, wired and wireless high-speed internet access, audio/visual equipment available, dedicated banquet and conference services staff, Hilton Guangzhou Tianhe is fully capable of handling the most elaborate and complex conventions, high-profile corporate meetings and wedding celebrations.

Completion date: August, 2011
Location: Guangzhou, China
Designer: City Group

Photographer: Hilton Guangzhou Tianhe
Area: 65,951m²

完成时间：2011 年 8 月
地点：中国，广州
设计师：城市组

摄影师：广州天河新天希尔顿酒店
面积：65,951 平方米

广州天河新天希尔顿酒店位于广州的中央商务区，距离白云国际机场仅有40分钟的车程，紧邻汇集机场地铁线、火车和汽车线路的交通枢纽中心。酒店的中心位置便于客人前往广州主要的购物和娱乐场所。客人能够在酒店附近享受广州最好的一切。

504间宽敞的客房和套房将提供最先进的设施和舒适服务，让客人的停留无忧无虑而又轻松无比。每个房间都采用了高品位的现代设计，落地玻璃窗将展示城市天际线无与伦比的风景。无线网络、高清电视和DVD播放机将为宾客们提供健康的创新娱乐体验。

酒店的六家餐厅和酒吧将提供各式各样的美味佳肴。随轩中餐厅拥有13间就餐包房，提供最精致的广东美食。意畔餐厅设有两间包房，在现代时尚的背景中供应正宗意大利美食。无贰全日制西餐厅将提供特色早午晚自助餐。独壹酒吧是会客会友的好去处，晚上有现场音乐表演。天河廊提供精致的下午茶和甜点，适合朋友休闲小聚。林下轩让人在轻松别致的露天环境中品鉴西班牙小菜、本地小食及沁人心脾的饮品。

1,900平方米的多功能会议和宴会空间配有有线和无线高速网络接口、音频/视频设备以及专业的宴会服务人员。广州天河希尔顿酒店能够举办最精致复杂的会议、高档公司会议和婚礼庆典。

1. Il Ponte' presents authentic Italian cuisine in a modern and chic setting.
2. Executive Lounge is furnished with numerous bubble-like pendants, as if the lounge is under the sea.
3. With traditional and classic Chinese decoration, the private dining room of Sui Xuan is added more elegance.

1. 意畔餐厅在现代时尚的背景中提供正宗的意大利美食
2. 行政酒廊装饰着不计其数的泡泡吊灯，让人感觉仿佛置身于海底
3. 随选中餐厅的包房装潢古香古色，气质优雅

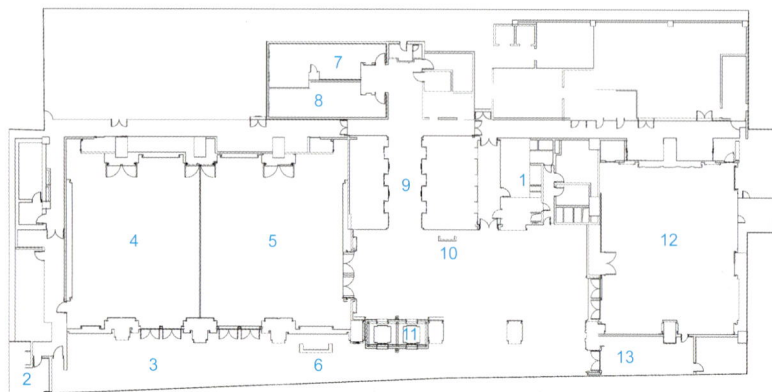

1. Staircase	1. 楼梯
2. Bride room	2. 新娘化妆间
3. Ballroom foyer	3. 大宴会厅前厅
4. Ballroom A	4. 大宴会厅A
5. Ballroom B	5. 大宴会厅B
6. Service station	6. 服务台
7. Women toilet	7. 女卫生间（公共）
8. Men toilet	8. 男卫生间（公共）
9. Lift lobby	9. 电梯厅
10. Bar	10. 水吧
11. View lift	11. 观光电梯
12. Meeting room M1	12. 会议室M1
13. VIP meeting room V1	13. 贵宾会议室V1

1. The Grand Ballroom features colourful pendant lights and carpets, adding more vitality to the space with white palette.
2. Every Suite has a king-sized Serenity Bed which features an exclusive Serta mattress, luxurious linen bedding and a decorative bed throw.
3. Experience complete luxury in the elegant bathroom.

1. 大宴会厅以多彩的吊灯和地毯为特色，白色为整个空间增添了活力
2. 每间套房都配有特大号的大床，以奢华的舒达床垫、亚麻床品和床盖为特色
3. 在优雅的浴室中体验极致奢华

1. Staircase
2. Multi-functional room M8
3. Multi-functional room M7
4. Multi-functional room M6
5. Multi-functional room M5
6. Women toilet
7. Men toilet
8. Lift
9. View lift
10. Meeting room M3
11. Meeting room M2
12. VIP meeting room V2

1. 楼梯
2. 多功能会议室M8
3. 多功能会议室M7
4. 多功能会议室M6
5. 多功能会议室M5
6. 女卫生间
7. 男卫生间
8. 电梯间
9. 观光电梯
10. 会议室M3
11. 会议室M2
12. 贵宾会议室V2

1. The Grand Ballroom features colourful pendant lights and carpets, adding more vitality to the space with white palette.

2. Every Suite has a king-sized Serenity Bed which features an exclusive Serta mattress, luxurious linen bedding and a decorative bed throw.

3. Experience complete luxury in the elegant bathroom.

1. 大宴会厅以多彩的吊灯和地毯为特色，白色为整个空间增添了活力

2. 每间套房都配有特大号的大床，以奢华的舒达床垫、亚麻床品和床盖为特色

3. 在优雅的浴室中体验极致奢华

1. Staircase	1. 楼梯
2. Multi-functional room M8	2. 多功能会议室M8
3. Multi-functional room M7	3. 多功能会议室M7
4. Multi-functional room M6	4. 多功能会议室M6
5. Multi-functional room M5	5. 多功能会议室M5
6. Women toilet	6. 女卫生间
7. Men toilet	7. 男卫生间
8. Lift	8. 电梯间
9. View lift	9. 观光电梯
10. Meeting room M3	10. 会议室M3
11. Meeting room M2	11. 会议室M2
12. VIP meeting room V2	12. 贵宾会议室V2

HILTON CHENNAI
金奈希尔顿酒店

A city of contrasts and diversities, Chennai is the forth-largest city in India. The cultural heritage of this city provides a rich backdrop for the gateway to India in the south. Traditional Tamil Hindu and European (including Portuguese, Dutch and British) influences coexist – here the past is able to live side by side with the present while industry looks to the future. The design captures this rich confluence of tradition, culture while paying homage to the Mid-Century Modern aesthetics that was so prevalent when Hilton Hotels started in the last century.

Upon entering, traditional Tamil carving inspired stone inlay pattern on the floor leads guests from the entrance through the circular lobby formed by beautifully contemporised Jali screen suspended from above. Guests arrive to a set of very stylised registration desks done in exotic stone. Serene pools of water on either side of the lobby provides a sense of respite for also the adjacent lounge-café, which is also designed to evoke a sense of luxury and elegance from a past era while providing a fresh new attitude. A distinct sense of place and timelessness is conveyed in the space and it activates the ground floor as a place to linger.

Similar Mid-Century revival approach is adopted throughout all spaces from signage design to the ballroom chandelier. This is especially evident in the typical guestroom design as the design team custom-designed every piece of furniture and furnishings to strike the perfect balance between the past and present to create the most sophisticated and luxurious room. The artwork over the bed was conceived as part of the interior architecture to inject a sense of place. It integrates a mirror with a panel upholstered by a variety of randomly selected local silk stitched together so it would appear differently in every room.

The spa level is a tranquil respite that the guest is drawn to from the moment they reach the lift lobby. Wood planking lined with polished white river stones mark the departure of the spa and fitness experience. The entry is lined with frosted glass stepped wall sconces. Locker rooms with yellow onyx accents provide sauna and steam as well as small plunge pools. Private treatment rooms and outdoor relaxation areas with soothing fountain are additional amenities provided for the guests. The roof terrace also creates an opportunity for an outdoor juice bar and private cabanas which line the pool area. This space can be an active lounge that will be relaxing during the day as an oasis as well as an exciting candlelit roof top lounge in the evenings.

Overall the Hilton Chennai will be a new mark for hospitality in the region – showing respect to the unique location in the crossroads while capturing the modern freshness that the hospitality industry has pushed to the forefront with design.

Completion date: 2011
Location: Chennai, India
Designer: DiLeonardo (except restaurants)

Photographer: VRX Photography
Area: 14,531m²

完成时间：2011 年
项目地点：印度，金奈
设计师：迪里奥纳多（不包括餐厅部分）

摄影师：VRX 摄影
面积：14,531 平方米

作为印度第四大城市，金奈充满了反差和多样性。城市的文化遗产为印度南端的门户提供了丰富的历史背景。传统的泰米尔印度人和欧洲人（包括葡萄牙人、荷兰人和英国人）的影响在这里共存，历史与现在能够和平相处，并且共同面向未来。设计抓住了这种丰富的传统与文化的汇合，同时也对希尔顿酒店创始时期的中世纪现代美学表达了敬意。

一走进酒店，地板上从传统泰米尔雕刻风格的石刻图案便引领着客人从入口走到了圆形酒店大堂。客人将到达一系列时尚的异国情调石制登记台。大堂两侧的幽静水池为紧邻的休闲咖啡吧提供了舒缓的氛围。咖啡吧的设计令人想起了旧时代的奢华，同时又提供了全新的生活态度。空间传递出独特的地方感和永恒感，让酒店一楼变成了消磨时光的好去处。

类似的中世纪复兴设计散步在整个空间，从引导标示到宴会厅的吊灯，无所不在。这在标准客房的设计中尤为明显，设计团队特别定制了每样家具和室内城市，力求在过去和现在之间形成完美的平衡，打造出最精致奢华的房间。床头的艺术品是室内设计的一部分，突出了地方感。各种各样的随机挑选的本地丝绸被拼接在一起，形成了衬垫面板，与镜子一起形成了每个房间独一无二的装饰品。

水疗中心是一个宁静之所，客人们会被它深深吸引。两侧铺着白色雨花石的木板标志着水疗和健身体验的开始。入口处排列着毛玻璃阶梯壁灯。采用黄玛瑙装饰的更衣室将提供桑拿、蒸汽浴室和小型跌水池。私人治疗室和配有喷泉的露天休闲空间是附加的休闲设施。屋顶平台上设有露天果汁吧和沿着泳池的私人更衣室。这一空间可以作为活跃的场所：白天作为日光浴的好去处，晚上则能够举办屋顶烛光活动。

金奈希尔顿酒店将成为当地酒店业的新标志——既体现了对其独特地理位置的尊重，又捕捉到了酒店业的现代新鲜感，走在了设计的前沿。

1. Exterior view of the hotel
2. Upon entering, traditional Tamil carving inspired stone inlay pattern on the floor leads guests from the entrance through the circular lobby.
3. The roof terrace also creates an opportunity for an outdoor juice bar and private cabanas which line the pool area.

1. 酒店外景
2. 走进酒店，地面上的传统泰米尔雕刻图案引领着客人穿过圆形酒店大堂
3. 屋顶阳台的游泳池旁设有露天的水吧和私人更衣室

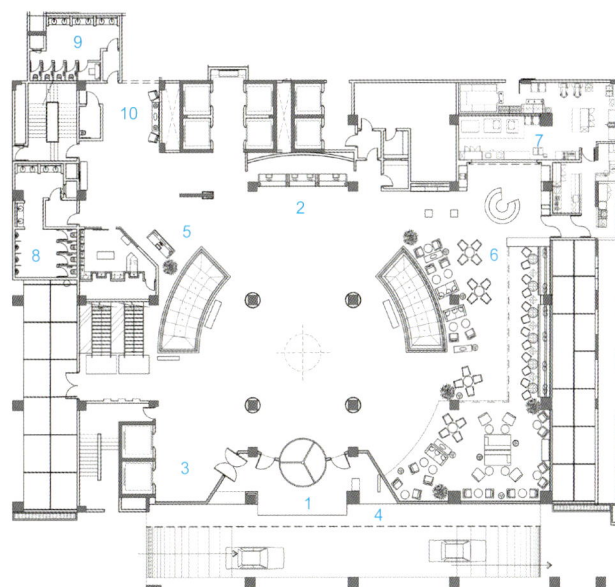

1. Entry
2. Reception
3. Lift lobby
4. Bellman station
5. Concierge
6. Café lounge
7. Bakery open kitchen
8. Male restroom
9. Female restroom
10. Seating

1. 入口
2. 前台
3. 电梯大厅
4. 行李服务台
5. 门房
6. 咖啡厅
7. 面包房开放厨房
8. 男洗手间
9. 女洗手间
10. 休息区

1. The classroom set-up in ballroom
2. Stay in an Executive Room for exclusive access to the Executive Lounge, which offers incredible city views and a relaxed ambience.
3. Every piece of furniture and furnishings strikes the perfect balance between the past and present to create the most sophisticated and luxurious room.

1. 宴会厅的会议布置
2. 由行政房可以进入行政酒廊，享受非凡的城市美景和放松的氛围
3. 每件家具和装饰都在过去和现在之间达成了完美的平衡，打造出最精致奢华的客房

1. Unwind in this comfortable Twin Guest Room decorated in a fusion of traditional and contemporary styles.
2. The artwork over the bed was conceived as part of the interior architecture to inject a sense of place. It integrates a mirror with a panel upholstered by a variety of randomly selected local silk stitched together so it would appear differently in every room.
3. The elegant bathroom in the guest room

1. 舒适的双人客房混搭传统与现代两种装饰风格
2. 床头的艺术品是室内设计的一部分，打造了空间感。它将镜子和拼接的本地丝绸结合在一起，随机挑选的布料让每间房间都各有不同
3. 客房中优雅的浴室

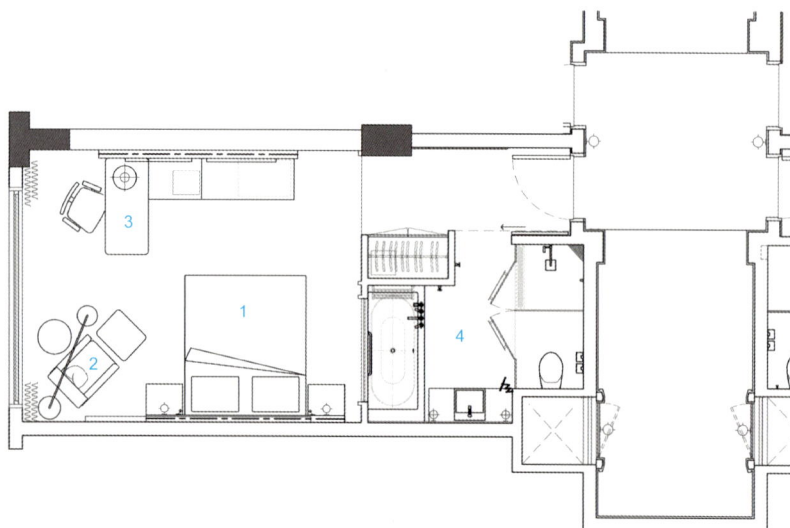

1. Bedroom
2. Seating area
3. Study area
4. Bathroom

1. 卧室
2. 起居区
3. 工作区
4. 浴室

WALDORF-ASTORIA SHANGHAI ON THE BUND

上海外滩华尔道夫酒店

The challenge for the designers is clear. The Waldorf-Astoria on the Bund, which will open in the former Shanghai Club, is tasked with taking its place among the world's finest hotels. However, as opposed to cities like Paris, London or New York where hotels in landmark properties are an intrinsic part of the city's hospitality map, Shanghai has little experience of managing restoration projects.

Adding to the challenge is the fact that expectations of travellers to China have heightened over the last five years. The opening of international-standard hotels in major Chinese cities has elevated demands for comfort, hospitality and service. Shanghai's own breakneck development has become associated with modernity and futurism, so updating these landmark buildings without compromising their architectural traditions means navigating uncharted waters.

The objective, however, is to create interior spaces that update each building's unique personality. For while there is a consensus worldwide that luxury travellers like to stay in hotels located in older buildings blending local traditions and history, the hotel will operate in a very competitive marketplace. Over the next two years, Shanghai will welcome at least ten other luxury hotel openings.

The guest rooms and restaurants are vital components for landmark hotel. As part of the design process, the designers engaged as many references as possible and followed every lead they could find. The designers watched black-and-white movie reels shot inside the building, and even ran a campaign for people to share their memories of this magnificent building. They also assessed how dining expectations are evolving in Shanghai, and looked at the best restaurants worldwide, both freestanding and in hotels, to determine the right elements for attracting not only hotel guests but also a cross-section of Shanghai society.

Local design elements will be incorporated sparingly, including artworks hanging in public spaces and design motifs in the bedrooms. It is important not to create a generic interior that references only the building's heritage or location. However, the project has an objective: to become world-class hotels that balance historic appeal with contemporary luxury — and which contribute to the regeneration of The Bund.

Completion date: September, 2010
Location: Shanghai, China

Designer: Ian Carr and Connie Puar (HBA)
Photographer: Ken Hayden

完成时间：2010 年 9 月
地点：中国，上海

设计师：伊恩·卡尔和科尼·布瓦尔（HBA）
摄影师：肯·海顿

设计师所面临的挑战十分明确：开在前上海总会的外滩华尔道夫酒店的目标是跻身世界顶级酒店之列。但是与巴黎、伦敦和纽约不同，它们的酒店是城市酒店业地图的一部分，是地标性建筑，而上海则对修复项目几乎没有什么经验。

近五年以来，来到中国的游客的期望值不断提高。中国主要城市中国际级酒店的开张提升了人们对舒适、好客以及服务的需求。上海的飞速发展充满了现代化和未来元素，因此，对这些地标性建筑的保护性改造升级意味着在未知的水域航行。

然而，设计的目标是打造能够提升建筑特色的室内空间。由于奢华型旅客更倾向于入住在古老建筑中而富有当地传统和历史的酒店，华尔道夫酒店将在竞争激烈的市场环境中运营。在未来的两年内，上海将迎来至少十家全新的奢华酒店。

客房和餐厅是酒店的重要组成部分。作为设计流程的一部分，设计师参考了许多资料并且利用了一切可以找到的线索。设计师在建筑内观看黑白影片，甚至专门举办了建筑记忆分享活动。他们还研究了人们对上海的餐饮期待，参考全球各地的餐厅，以此来决定既能够吸引酒店宾客又能够立足上海社交界的餐饮元素。

设计将保守的采用本地设计元素，其中包括公共空间悬挂的艺术品和卧室中的设计图案。酒店仅靠参考建筑历史和地理位置来进行室内设计。然而，酒店的目标是：成为均衡历史和现代奢华的世界级酒店，同时为外滩的复兴做出贡献。

1. In Grand Brasserie, lavish marble columns and countertops, polished wood tables and cushioned lounges and banquettes re-create a modern Manhattan brasserie setting.
2. One of private dining rooms at Grand Brasseire provides a sense of peaceful retreat in urban Shanghai.
3. Another private dining room at Grand Brasseire

1. 百味园里，奢华的大理石柱和桌面、抛光木桌和柔软的座椅重现了现代曼哈顿餐厅风情
2. 百味园的包房在上海打造了一个平静的休息空间
3. 百味园的另一间包房

1. Peacock Alley is an upscale lounge that orchestrates the perfect symphony of Waldorf-Astoria's classic surroundings and fresh contemporary ambiance.
2. Decorated in timeless style, library lounge reflects the Waldorf's legendary residential ambience.

1. 羿庭是一家高端酒廊，精心协调了华尔道夫经典布置和现代气息
2. 图书室采用了经典风格的装饰，反映了华尔道夫传奇的居家氛围

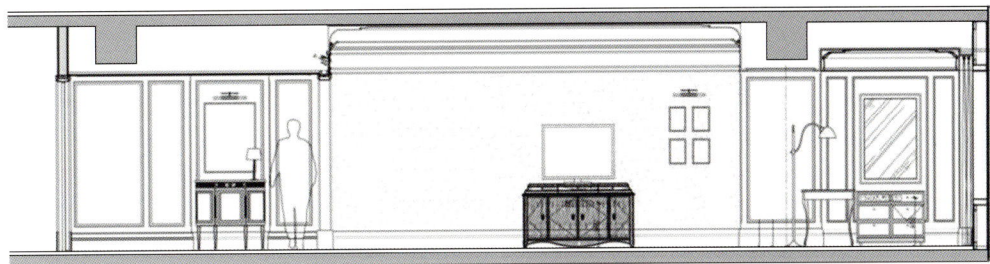

1. Local design elements are incorporated sparingly in the living room of Presidential Suite, including artworks and furnishings.

2. Luxury River Suite owns a fresh colour palette with glimmering chandelier, which forms a kind of contemporary luxury.

3. Fitted with TV in the mirror and plush amenities, "better than home" haven completes here.

4. Flower motif is throughout Luxury Suite, which refreshes and unwinds the tired mood.

1. 总统套房的客厅里展示了少量本地设计元素，包括艺术品和室内陈设

2. 奢华河景套房拥有新鲜的色彩，配有闪烁的吊灯，形成了现代奢华感

3. 豪华的设施和电器让客房变成了"比家更好的地方"

4. 奢华套房里满是花朵图案，让疲惫的身心得到了放松

HYATT HOTELS AND RESORTS
The Pioneer of Modern Hotel Architecture

凯悦酒店集团——现代酒店建筑开拓者

In 1967, Hyatt Hotels and Resorts built the first "Atrium Hotel" in Atlanta, opening up the age of hotels with magnificent central lobbies. Hyatt Hotels and Resorts therefore was recognised by the world and developed rapidly. Today, Hyatt worldwide portfolio consists of nearly 800 hotels. Hyatt has carried on the brand's excellent heritage in their architectural characteristics.

In architectural design, Hyatt combines elegance and luxury perfectly and blends classics and innovations thoroughly. Hyatt hotels all own their unique façades: pyramid, trapezoid, arch, and cylinder, all presenting a breath of fresh air. In interior design, Hyatt hotels prefer magnificent and distinctive entrance halls and lobbies. In addition, Hyatt hotels reflect local features everywhere and provide guests with local accents.

Recently, featuring their distinct exterior and interior designs, Park Hyatt, Grand Hyatt and Hyatt Regency are leading Hyatt Hotels and Resorts.

Compared with Hyatt Regency and Grand Hyatt, Park Hyatt aims to create deluxe boutique hotels in pursuit of personalised experience and European classic style. In architectural design, Park Hyatt integrates classic flavours into contemporary architecture to achieve a balance and unity between tradition and modern. In the term of interior design, Park Hyatt focuses on details and highlights serenity through simple yet exquisite design methods. The renowned art collection in Park Hyatt enable guests to experience classics even without notice. Besides, the location of a Park Hyatt hotel is always well considered. Located in the centre of a modern city, Park Hyatt enjoys a privileged position. Overlooking from the higher levels of the hotel, guests can either enjoy a city view or listen to the hustle and bustle of the city.

As a large luxurious business hotel brand of Hyatt, Grand Hyatt always selects its site in an international city or resort. Through its rich and diversified facilities, Grand Hyatt is a custom-designed leisure space for business clients. A Grand Hyatt hotel always possesses a magnificent exterior to highlight guests' eminent statuses. In interior design, Grand Hyatt pays particular attention to open space, reasonable furniture arrangement, magnificent lobby, comfortable guestrooms and gorgeous conference rooms. With its signature entrance hall, Grand Hyatt provides excellent conditions for business dinners and conferences.

As the core brand of Hyatt Hotels and Resorts, Hyatt Regency is a group of medium luxurious business hotels. Hyatt Regency has an innovative spirit in both architectural and interior designs, directly inheriting the architectural spirit of Hyatt Hotels and Resorts.

This chapter selects ten hotels of these three brands, exhibiting the excellent designs of Hyatt Hotels and Resorts.

1967年，凯悦酒店集团在亚特兰大建成了世界第一家"中庭酒店"，从此开创了在奢华酒店设计中建造中空状、宏伟气魄大堂的先河。凯悦酒店凭此享誉世界，飞速发展，至今集团旗下已有奢华酒店近800家，此后，凯悦酒店集团更将注重建筑特色的优良传统传承至今。

在建筑设计方面，凯悦酒店集团将典雅与奢华完美结合，将经典与创新融会贯通。旗下酒店均拥有独特的建筑立面：金字塔形、梯形、拱形、圆柱形……往往给人耳目一新之感。而在室内设计方面，凯悦酒店集团钟爱气势宏伟，风格独特的门廊、大堂，另外，在酒店各处会体现各地的本土特色，通常令客人足不出户就能感受到当地的风土人情。

近年来，凯悦酒店集团旗下的柏悦酒店、君悦酒店、凯悦酒店均以鲜明的室内外建筑设计特色领跑其旗下各酒店品牌。

相比凯悦、君悦酒店，柏悦酒店为追求个性化体验和欧洲典雅风格的房客精心打造了超豪华型精品酒店。在建筑外部空间设计上，它将古典气息融入现代建筑风格，实现了传统与现代的平衡与统一。在室内设计上，柏悦酒店注重细节，凭借简约精致的设计手法彰显着惬意的宁静。柏悦酒店内典藏的艺术品更令房客在不经意间品味经典。此外，柏悦酒店的选址也颇为讲究。居于现代都市的中心，柏悦酒店享有优越的地理优势。从酒店高层俯瞰，房客们或饱览都市流光溢彩，或聆听都市脉动的喧嚣熙攘。

作为凯悦酒店集团旗下的大型豪华商务酒店品牌，君悦常选址于国际化都市或度假胜地。凭借丰富多元的酒店设施，君悦酒店为商务人士量身打造了休闲娱乐空间。君悦酒店通常拥有气势恢宏的建筑外观，以彰显房客的显赫地位。室内设计上，君悦酒店讲究空间宽敞、家具布置合理、大堂气魄堂皇、客房温馨舒适、会议室华丽尊贵，加之凯悦酒店集团标志性的门廊大厅，君悦酒店为商务晚宴和会议的召开提供了极佳的便利条件。

凯悦酒店是凯悦集团的核心品牌，属中型豪华商务酒店。凯悦酒店无论在建筑和室内设计方面都独具创新精神，是凯悦酒店集团建筑精神的直接承接品牌。

本章选取隶属于这三大品牌酒店的十个酒店项目，层层揭开凯悦集团酒店的设计风采。

HYATT REGENCY DUSSELDORF

杜塞尔多夫凯悦酒店

The Hyatt Regency Dusseldorf, set to act as focal point and destination on the tip of the island in the Media Hafen Dusseldorf. On the site of the former Monkey Beach an iconic design has been created, linked to the city by a pedestrian bridge. The area, with a stunning view on the Rheinturm, contains some of the most spectacular modern architecture in Germany and is home to many fashion houses, design- and advertising agencies. From a distance the bar pavilion Pebble's, situated on the plateau above the public spaces, will be the shining recognition point for all visitors.

The completely glazed ground floor of the hotel entices guests showing the different aspects of the hotel. The interior is based on emphasising this magnificent view and on creating warmth and intimacy in a modern building. A recurring motif in the interior design of the hotel, which motif was designed by FG stijl, are water reeds referring to the island environment.

The black Norwegian slate floor at the reception area creates the perfect background for the most dominant aspect of the public spaces, the Golden Box.

The wooden screens on the outside emphasise the library-like feeling inside, creating privacy and a warm atmosphere. A stunning artwork of fashion designer Wolfgang Joop makes it all complete. The orchid pattern carpet was designed by FG stijl.

A central catwalk, designed by FG stijl, connects the three main areas of Dox Restaurant, inviting the ladies from Dusseldorf to parade on their high heels. Guests are drawn towards the elegant free floating curved staircase at the end of the area, designed to dominate but also to allow guests the chance to see and be seen. Up the stairs is Pebble's, the hotspot for after dinner drinks.

The jewel of the hotel is the bar pavilion aptly named Pebble's. This shimmering river bolder shaped pavilion of polished stainless steel is set to shine as a central feature on the tip of the island. The large roof terraces around Pebble's enable guests to enjoy the view of the entire harbour form above to the water. The shimmer and shine of the outside is also to be found inside the bar, especially in the mosaic tiles on the floor, which is turned into a mere jewel when the sun shines on it.

The Ballrooms and meeting rooms are accessed either from the lobby of the hotel or through the "movie star entrance", the central section in the stairs between the two buildings. A seven-metre-long special commissioned art work by Sal attracts the visitor towards this main area. In the almost 500-square-metre ballroom, the large glass ceiling is covered with water. The movement of the water above turns the especially designed carpet of the ballroom

Completion date: February, 2011
Location: Dusseldorf, Germany
Designer: Sop architekten, FG stijl

Photographer: Peter Peirce, Chris Taggart
Area: 4,738m²

完成时间：2011 年 2 月
项目地点：德国，杜塞尔多夫
设计师：索普建筑事务所；FG 风格工作室

摄影师：彼得·皮尔斯；克里斯·塔格特
面积：4,738 平方米

into a water lily pond. Looking up guests can see the two towers above shimmering through, creating a wonderful city feeling.

The spa and gym are located on the ground floor. The gym has an excellent and inspiring view on the pedestrian bridge connecting the island to the city, which is lit at night.

The spa area contains five multi-functional beauty- and treatment rooms including one double room and a Vichy treatment room. The focal point in each treatment room is the mosaic wall of natural coconut, creating a warm surrounding. In each treatment room a purple amethyst-crystal is placed centrally. The light and airy relaxation area of the spa has a terrace overlooking the water and the shimmering Pebble's pavilion.

Descending the stone staircase leads to the hidden treasure: a sunken natural stone bath, with a golden floor and a view on the inner garden.

The Regency Club situated on the 17th floor has floor-to-ceiling windows, creating a true retreat above the hustle and bustle of the city. The custom woven rug with a pattern of a river viewed from above, connects the different areas, among which a meeting room which can be closed off with glass doors covered with steel reeds.

The hotel contains 303 guestrooms, including 260 king rooms, 30 twin rooms, 10 junior suites, 2 executive suites and 1 presidential suite.

One of the biggest challenges was the unusual structure of the building because of the cantilevered design. Throughout the building, enormous diagonal concrete pillars have been installed. It was exciting to create an elegant but practical interior design, resulting in surprising rooms.

The standard king room is not very standard. In this new concept the first half of the room is designed as a luxurious dressing and pampering area. Behind this area in the second half of the room, the bed is facing the window to enjoy the spectacular view. This view can be seen from the bath tub as well, because of the glass between the bath tub and the bedroom.

In the suites the view was the focal point for the design: whether a guest is seated in the living area, in bed, in the bath or in front of the make-up table, the view over the water and the city is spectacular.

The presidential suite is complete with grand piano, double fireplace and the best views in town positioned as it is on the entire width of the building facing the pedestrian bridge.

The inside of the lifts, the carpet in the corridors and even the air grilles: everything was especially designed by FG stijl, with reeds and nature as recurring motifs.

1. The entrance to the hotel is under the enormous cantilever canopy created by the rooms above.
2. The view inside the entrance of the hotel

1. 酒店入口上方是客房所形成的巨大穹顶
2. 酒店入口内部

1. The counter in reception area
2. Guests are drawn towards the elegant free floating curved staircase at the end of the area, designed to dominate but also to allow guests the chance to see and be seen.
3. Just like the outside of Pebble's, the décor inside is also shining and stunning.

1. 大堂接待处
2. 末端的悬浮型弧形楼梯在设计中处在支配性地位，让宾客们相互认识
3. 正如卵石酒吧的外观一样，它的内部装饰同样闪闪发亮

1. Executive boardroom
2. Meeting room
3. Ballroom
4. Hotel lobby
5. Lounge
6. Dox Bar
7. Dox Restaurant
8. Sushi Bar
9. River Salon
10. Café D

1. 首席会议室
2. 会议室
3. 宴会厅
4. 酒店大堂
5. 休息区
6. Dox酒吧
7. Dox餐厅
8. 寿司吧
9. 沙龙
10. 咖啡厅D

杜塞尔多夫凯悦酒店是这个米提亚港口小岛尖端的聚焦景点。酒店坐落在前猴子海滩，通过步行桥与城市相连。酒店所在的区域享有莱茵塔迷人的景色，远眺着一些德国最壮丽的现代建筑并且坐落着许多时装店、设计公司和广告公司。从远处看，高地上的卵石酒吧将是酒店的闪光点。

酒店的一楼全部采用了玻璃幕墙，吸引着宾客进入酒店内部探索。室内设计以壮丽的视野和温馨私密的现代感为特色。酒店室内设计中重复出现的芦苇主题由FG风格工作室设计，反映了小岛的自然环境。

前台接待区的黑色挪威石板地面为公共区域——金色盒子——提供了完美的背景。休息室外面的木制屏风凸显了内部的图书馆感觉，营造出私密而温馨的氛围。由时装设计师沃尔夫冈·乔普所设计的艺术品令人眼前一亮。兰花图案的地毯则由FG风格工作室设计。

同样由FG所设计的中央天桥连接了多克斯餐厅的三个区域，让女士们踩着高跟鞋自信前行。末端的悬浮型弧形楼梯在设计中处在支配性地位，让宾客们相互认识。楼梯上方的卵石酒吧吸引着人们在晚饭后去喝一杯。

卵石酒吧宛如酒店顶部的明珠。这个闪闪发亮的卵石形酒吧采用了抛光不锈钢材料，是小岛尖端的中央景观。环绕酒吧四周的巨型屋顶平台让宾客们可以尽享整个港口的景色。酒吧外部的闪光感在室内也得到了体现，地面上的马赛克地砖在阳光下如宝石般闪亮。

宴会厅和会议室可以从酒店大堂和"明星入口"——两座建筑楼梯之间的中心街面——两侧进入。由萨尔特别制作的艺术品引领着宾客进入这一区域。在近50米的宴会厅中，巨大的玻璃天花板上方全是水。上方水流的运动将特别定制的地毯变成了莲花池。宾客们能够看到宴会厅上方闪耀的大楼，营造出奇妙的都市感。

水疗中心和健身房设在一楼。连接小岛和城市的天桥上在夜晚被点亮，为健身房提供了非凡而壮丽的景观。

水疗区包含5个多功能美容治疗室，其中有一间双人房和一间薇姿治疗室。治疗室设计聚焦于天然椰子马赛克墙面，营造出温馨的氛围。每个治疗室的中央都摆放着一座紫水晶。轻盈而放松的水疗区拥有一个俯瞰水面和卵石酒吧的平台。

石阶向下通往秘密宝藏：一个下沉式天然石浴池，浴池配有金色的地面，可以眺望室内花园。

18楼的摄政俱乐部设有落地窗，在喧嚣繁忙的都市中营造了一片世外桃源。特别定制的编织地毯以河景为图案，连接了各个不同的区域，其中一间会议室的大门上雕刻了钢铁芦苇。

酒店拥有303套客房，其中包括260套国王房、30套双子房、10套普通套房、2套行政套房和1套总统套房。

建筑的悬臂式设计为客房设计提出了重大的挑战。大量对角线混凝土柱遍布整个大楼。优雅而实用的室内设计打造了令人惊喜的房间。

标准国王房其实并不标准。客房的一半被设计成豪华的穿衣护理区。另一半的大床朝向窗户，享有壮丽的景色。由于浴缸和卧室通过玻璃隔开，从浴缸里也能向外眺望。

套房的视野是设计的焦点：无论宾客是在起居区、床上、浴室还是梳妆台前，水上和城市的视野都十分令人惊叹。

总统套房配有大钢琴、双层壁炉和最好的视野，套房的整面墙都面对着步行桥。

电梯内部、走廊地毯乃至通风花窗，所有细节都由FG风格工作室特别设计，芦苇和自然景观是不变的主题。

1-2. The lounge emphasises the library-like feeling inside, creating privacy and a warm atmosphere. The fire place makes the lounge feel like home.

1.2. 休息室外部的木制屏风凸显了私密而温馨的氛围

1. A spectacular entrance to the ballroom is set to be the pre-function area before many events.
2. In the 500-square-metre ballroom, the large glass ceiling is covered with water. The movement of the water above turns the especially designed carpet of the ballroom into a water lily pond.

1. 宴会厅华丽的入口处是活动准备区
2. 在近500平方米的宴会厅中，巨大的玻璃天花板上方全是水；上方水流的运动将特别定制的地毯变成了莲花池

1. The layout of living room in suite
2. The living room of the presidential suite is complete with grand piano and double fireplace.
3. The bed in standard king room is facing the window to enjoy the spectacular view.

1. 套房客厅的布局
2. 总统套房的客厅配有大钢琴和双层壁炉
3. 标准国王房的大床朝向窗口，享受着壮丽的视野

GRAND HYATT MACAU

澳门君悦酒店

Grand Hyatt Macau comprises two wave-inspired towers within City of Dreams, the aquatic-themed, integrated entertainment resort located on Cotai.

Approaching Grand Hyatt Macau, guests are greeted by hotel hosts and guided into the vast lobby, with its staggering 22-metre-high ceiling. Against a sweeping backdrop of travertine inlaid with red rainforest marble; overlapping geometric wall panels; hexagonal columns; and black marble flooring, a striking sculpture by leading Chinese artist Danny Lee dominates the grand space. Water cascades over a giant stainless-steel hemisphere, while above it, curved stainless-steel ribbons shower down from an illuminated cloud motif in the ceiling.

Above the entrance area, the entire second level of the hotel is dedicated to event space. The colossal, pillar-less Grand Ballroom with its eight-metre-high ceiling covers 1,911 square metres and can cater up to 2,000 theatre-style and 1,300 for a banquet. Next door, Salão do Teatro measuring 774 square metres, is equally impressive. Translating as "show theatre" in Portuguese, Salão do Teatro is the first event space of its kind, featuring an open show kitchen capable of accommodating up to 20 chefs at a time, with video cameras beaming live culinary action to a pair of projectors. The pillar-less ballroom sports a 6.5-metre-high ceiling and its own control room, and can cater up to 360 for a banquet. Uniquely, it also has its own private pre-function area, which can cater up to

300 for cocktails, and is the only ballroom in Macau blessed with natural daylight.

For its two main restaurants, Grand Hyatt Macau has been inspired by two award-winning eateries in its sister hotels. Beijing Kitchen on the Lobby replicates the highly successful format of Grand Hyatt Beijing's Made in China restaurant, and is one of the signature restaurants at City of Dreams. Meanwhile, mezza9 Macau on Level 3 has been inspired by the famous mezza9 at Grand Hyatt Singapore and focuses on international fare. Created by the cutting-edge, Japanese interior design firm SuperPotato, the 292-seater space is defined by giant roughly-hewn granite blocks creating counters and lining walls, as well as an array of lattice-patterned, metal screens.

Measuring 64 square metres, "Grand Suites" in the Grand Tower feel particularly spacious and feature an entirely separate living area, divided by a sliding door. Floor-to-ceiling picture windows provide uninterrupted views of Cotai or the south bank of the Pearl River and all rooms feature a king-size bed or two double beds; images of Macau by British photographer William Furniss; and a contemporary marble bathroom with a deep-soaking tub, twin vanity areas, a wet room and a glass-walled rain-shower. Over in the Grand Club Tower, the "Grand Deluxe" rooms measure 52 square metres and offer an additional freestanding elliptical bath looking out over Cotai or the Pearl River; a walk-in dressing room;

Completion date: September, 2009
Location: Macau, China
Size: 136,641m2

Designer: HBA
Photographer: Samuel Nugroho

完成时间：2009 年 9 月
项目地点：中国，澳门
面积：136,641 平方米

设计师：HBA
摄影师：萨缪尔·纳格洛霍

1. At the heart of the hotel is the main lobby.
2. The lobby lounge is highlighted by a striking cluster of copper pendant art pieces.

1. 酒店大堂
2. 酒店大堂内的艺术品引人注目

original, nature-inspired art by the Australian painter Denis Murrell. On Level 35 of the Grand Club Tower is the Chairman Suite; a 275-square-metre private residence with panoramic city views, which offers a security-monitored entrance; a spa treatment area; a private gym; a kitchen with separate seating; a dining area; a work space; and a comprehensive home theatre sound system. Ideal for private entertaining, the suite also features a separate entertainment quarter with a pre-dining area; a formal 10-seater dining room; butler service; and a second entrance for service staff.

澳门君悦酒店由两座波浪形的双子楼组成，位于路氹地区的水上主题度假村——新濠天地内。

澳门君悦酒店坐落于路氹以水为主题的综合娱乐度假胜地——新濠天地。酒店由两幢呈波浪形的主楼组成。一进酒店大堂，就会被其豪华恢宏的环境氛围所感染。大堂以镶有红色热带雨林大理石的石灰华作优美背景，配以纵横交错的几何图案墙身、六角形大支柱、黑色大理石地板，充分凸现着著名华裔艺术家李展辉先生设计的触目雕塑，以此展现敬畏大自然的主题。丝带状的不锈钢垂坠装饰从绘有彩云图案的天花板上吊下，俨如坠落的雨水。在其下方置有巨型不锈钢流水地球仪，灵动地表现出大自然生生不息的本质。对客人来说，耀眼夺目的装置组合体现出酒店的设计美学和空间体验的品位。特高的酒店大堂楼底，加上大量玻璃装饰，结合从户外引入室内的天然光线，令酒店充满现代时尚气息，并创造出开阔的空间。

在门厅区域的上方，酒店的整个二层被设置成会议及展览场地。巨大的无柱式大宴会厅高8米，总面积可达1,911平方米，采用剧院式布局可以容纳2,000人，宴会式布局则可容纳1,300人。隔邻的盛会厅总面积774平方米，同样壮观。盛会厅是澳门第一家以开放式展示厨房为特色的活动空间，厨房可同时容纳20名大厨，摄像机会把烹饪现场直接转播到两台投影仪上。无柱式宴会厅高6.5米，配有独立的控制室，可同时招待360位宾客参加宴会。特别的是，它还拥有独立的私人活动前区域，可以为300人提供鸡尾酒，也是澳门唯一一间采用有天然光线的宴会厅。

澳门君悦酒店有两家重点餐厅，灵感来自姊妹酒店的两间得奖食肆。满堂彩餐厅复制了北京君悦酒店中餐厅的成功，是新濠天地中的标志性餐厅之一。三楼的mezza9 Macau餐厅从新加坡君悦酒店的mezza9餐厅中获得了灵感，专注于国际美食。这家由日本先锋室内设计公司SuperPotato所设计的餐厅可容纳292人，以巨大的天然岩石组成柜台和墙壁，配有栅格花纹的金属屏风。

64平方米的君悦套房显得异常宽敞，拥有独立客厅，与睡眠区由拉门隔开。落地风景窗享有路氹或珠江南岸的美景，配有特大双人床或两张单人床。墙上挂有英国摄影师William Furniss所摄的澳门景色。现代化的大理石浴室内设有浸浴缸、双梳妆台、湿室和雨淋式花洒浴间。嘉宾轩的豪华房面积达52平方米，拥有步入式化妆间和独立椭圆形浴缸，宾客可以在浴室中远眺路氹或珠江的美景，客房墙壁上则悬挂着澳大利亚画家Denis Murrell的艺术品。嘉宾轩的35楼是主席套房，275平方米的私人住所享有城市全景，入口处配有保安服务。套房配有水疗区、私人健身房、厨房、餐厅、工作区和家庭影院系统。套房的独立娱乐区配有餐前区、10坐席的餐厅、管家服务，并特设员工入口。

1. Grand Ballroom is set up with Opulence Ball Theme
2. Salão do Teatro is set up with Modern & Chic Theme
3. mezza9 Macau Main Dining

1. "濠门夜宴"主题
2. Modern & Chic主题
3. mezza9 Macau大厅

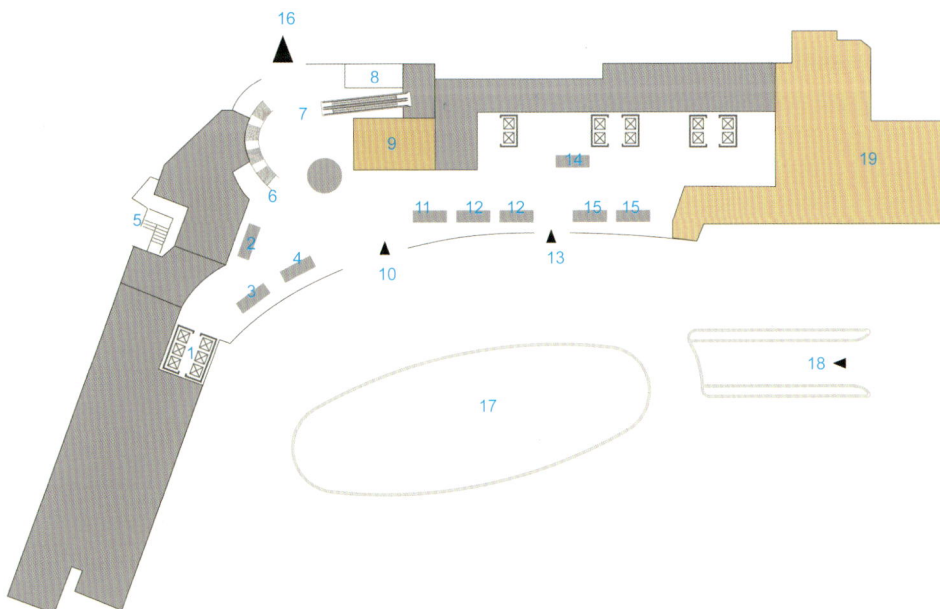

1. Guest lift	1. 宾客电梯
2. Grand club tower reception	2. 嘉宾轩楼接待处
3. Concierge	3. 礼宾部
4. Bell service	4. 行李部
5. Fire exit	5. 走火通道
6. Grand staircase	6. 楼梯
7. Escalator	7. 扶手电梯
8. Restroom	8. 洗手间
9. Lobby lounge	9. 大堂酒廊
10. Main entrance	10. 酒店正门入口
11. Gold passport elite counter	11. 凯悦金护照接待处
12. Grand tower reception	12. 君悦楼接待处
13. Entrance	13. 入口
14. Reception	14. 接待处
15. Check-out counter	15. 办理退房
16. The boulevard	16. 新濠大道
17. Fountain	17. 喷水池
18. Car park	18. 停车场
19. Beijing Kitchen	19. 满堂彩

1. Beijing Kitchen is decked out in rich Chinese décor accented.
2. The Grand Ballroom is divisible into four sections with a master control room, and it features catwalk lighting and advanced projectors.
3. Flexible meeting Salons, suitable for gatherings of between 40 and 100, also feature natural daylight.

1. 满堂彩餐厅采用了丰富的中式装饰
2. 大宴会厅分为四个部分，配有主控制室，拥有独特的天桥照明和先进的投影仪
3. 灵活的沙龙会议室适合40～100人会议，均有天然光线照射

1. The entrance of Isala Spa is decorated with a great amount of glass decorations.
2. The layout of Isala spa room
3. With separated living room and bar, Grand Suite King offers more space for entertainment.
4. Club Deluxe King offers contemporary décor, king bed and sculptured bath.

1. Isala Spa的入口处装饰着玻璃装饰品
2. 水疗中心理疗室
3. 君悦豪华客房拥有独立客厅和吧台，有更多的娱乐空间
4. 嘉宾轩客房配有现代装饰、双人床和雕花浴缸

HYATT REGENCY JING JIN CITY RESORT AND SPA

京津新城凯悦酒店

Hyatt Regency Jing Jin City Resort and Spa is the first all-day conference holiday resort in Northern China. Located in the Hot Spring City in Baodi District of Tianjin, the hotel is about 100 kilometres southeast to Beijing and 70 kilometres north to Tianjin, within one and a half hours' drive to both the cities. As an important component of a 104-square-kilometre project of Jing Jin City, Hyatt Regency Jing Jin City Resort and Spa has a magnificent design, matching up with the nobility of the royal palaces. The grand porch and fountain plaza, together with the innovative river landscape and exquisite water courtyard, have created an exotic feeling. The hotel is equipped with 793 luxurious guestrooms and suites, 4 international restaurants and bars and well-equipped conference rooms and exhibition area, which can accommodate hundreds of guests. The luxurious and fully-equipped recreational facilities include The Balneum Spa, YI Outdoor Hot Spring, 8-lane bowling alley, an indoor swimming pool, a gymnasium and Camp Hyatt for children. Together with the 27-hole standard international golf course, the hotel will create a diversified conference and holiday experience.

Completion date: September, 2007
Location: Tianjin, China
Area: 139,053m²

Designer: HEITZ PARSONS SADEK;
Jean-Philippe Heitz
Photographer: Samuel Nugroho

完成时间：2007年9月
项目地点：中国，天津
面积：139,053平方米

设计师：美国霍普斯有限公司；让－菲利普·黑兹
摄影师：萨缪尔·纳格洛霍

Completion date: September, 2007
Location: Tianjin, China
Area: 139,053m²

Designer: HEITZ PARSONS SADEK;
Jean-Philippe Heitz
Photographer: Samuel Nugroho

完成时间：2007年9月
项目地点：中国，天津
面积：139,053平方米

设计师：美国霍普斯有限公司；让－菲利普·黑兹
摄影师：萨缪尔·纳格洛霍

坐落于天津宝坻区温泉城的京津新城凯悦酒店是华北地区首家全天候会议度假酒店。它位于天津以北约70公里，北京东南约100公里处。作为占地104平方公里的京津新城整体发展项目的重要组成部分，京津新城凯悦酒店融合了古典皇宫建筑的尊贵气质，整体设计大气磅礴。伴以创意新颖的河流景观、典雅细腻的碧水庭园以及壮阔的门廊和喷泉广场，京津新城凯悦酒店更洋溢着异国风情。酒店设有豪华客房及套房793间、国际级餐厅酒吧4间、大型会议及展览场地功能齐备，可承办千人会议宴会。娱乐休闲方面，京津新城凯悦酒店设施完善豪华，包括温泉水疗Spa"道"、户外"逸"温泉、保龄球馆、壁球场、室内泳池、健身中心以及凯悦儿童营。此外，国标27洞高尔夫球场的建设又为入住京津新城凯悦酒店的各方宾朋提供了多元化及综合旅游度假的新体验。

1. Lakeview Lawn offers ideal backdrop for wedding venues.
2. The lobby resembles an ancient palace.

1. 湖景园是举办婚礼的理想场所
2. 酒店大堂宛如古代宫殿

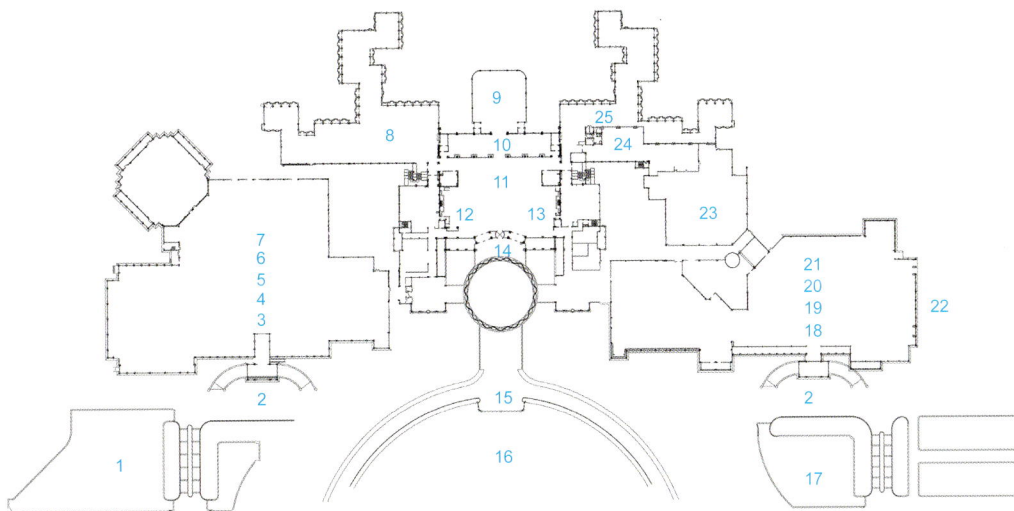

1. North parking area
2. Entrance
3. The Balneum Spa
4. Fitness centre
5. Bowling alley
6. Indoor swimming pool
7. Recreation Centre
8. Guestroom in North Wing
9. Glass House
10. Fountain Lounge
11. Lobby
12. Front office
13. Fountain Lounge Bar
14. Main entrance
15. Driveway
16. Big Fountain
17. South parking area
18. 23 Function rooms
19. The Ballroom
20. Regency Ballroom
21. Banquet and conference centre
22. Lakeview Lawn
23. The Garden-Chinese Restaurant
24. Ichiba Japanese Restaurant
25. Guestroom in South Wing

1. 北翼停车场
2. 入口
3. 水疗中心
4. 健身室
5. 保龄球馆
6. 室内游泳池
7. 娱乐及活动中心
8. 北翼客房
9. 水晶厨房
10. 碧泉茶园
11. 酒店大堂
12. 前台
13. 碧泉茶园酒吧
14. 正门
15. 酒店车道
16. 喷泉
17. 南翼停车场
18. 23间多功能厅
19. 大宴会厅
20. 凯悦宴会厅
21. 宴会会议中心
22. 湖景园-室外主题宴会
23. 8号紫园-中餐厅
24. 市集-日式餐厅
25. 南翼客房

1. The Ballroom features numerous impressive chandeliers and lotus lamps, of which decoration is full of Chinese traditional elements.

2. Integrated with other Chinese décor, the wall of Regency Ballroom decorated with Western paintings offers the impressive visual experience.

3. Fountain Lounge is a perfect place to enjoy the view from the lush gardens.

1. 大宴会厅装饰着无与伦比的吊灯和莲花座灯，充满了中国味道

2. 凯悦宴会厅墙壁上装饰着的西洋画打造了令人印象深刻的视觉体验

3. 碧泉茶园可坐享郁郁葱葱的花园美景

1. Guests not only can enjoy the authentic Chinese cuisine, but also are satisfied with Chinese traditional décor at The Garden Chinese restaurant.

2. Ichiba Japanese Restaurant offers a unique atmosphere similar to dining in a market village, featuring 17 open tatami-style private rooms.

3. Just like the name – Glass House, the buffet restaurant is surrounded by French windows.

1. 在8号紫园中餐厅宾客不仅可以享用地道的中餐，还能充分感受中国传统装饰

2. 市集——日式餐厅拥有17个榻榻米风格的包房及类似日式乡村市集的用餐氛围

3. 水晶厨房四周均采用法式落地窗设计

1. Chinese red spreads in the reception in The Balneum Spa.
2. The layout of The Balneum Spa treatment room
3. The Balneum Spa conservatory offers pure white atmosphere for echoing the theme of the spa – purity, balance and flow.
4. Modern, resort-style décor complements Regency Executive Suite.
5. The king room featuring one king bed with cotton linens and plush duvet offers comfortable and stylish atmosphere.

1. 中国红在"道"水疗的前台肆意伸展
2. "道"水疗理疗室的布局
3. "道"水疗纯白的氛围与水疗的主题相契合——纯粹、平衡、流畅
4. 现代度假风装饰了嘉宾轩行政套房
5. 大床房的床上铺有纯棉床单和豪华的羽绒被，舒适而时尚

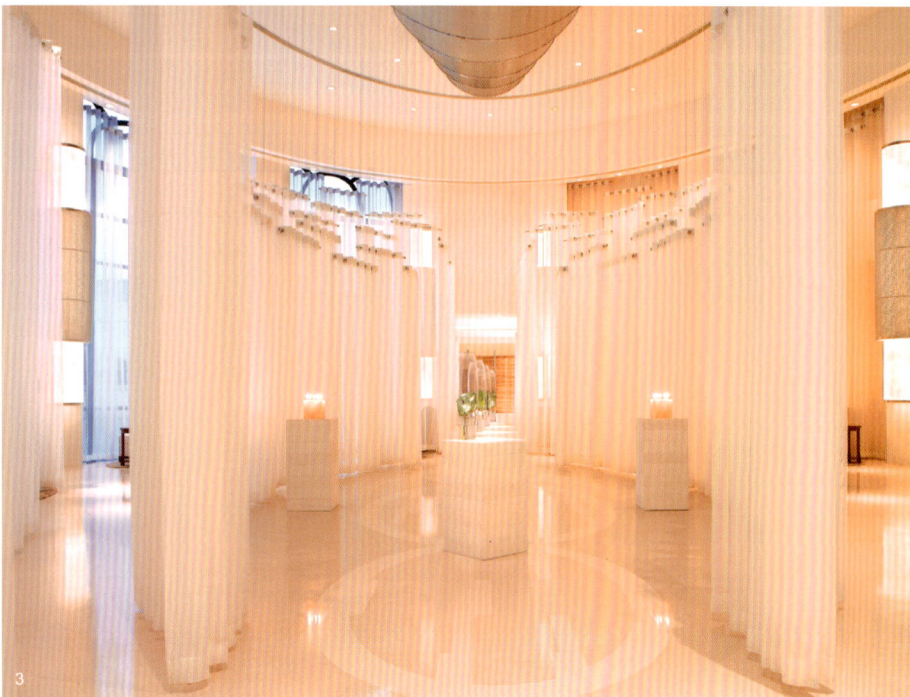

PARK HYATT SEOUL
首尔柏悦酒店

Set in the heart of Gangnam, the financial, business, shopping district of Seoul, the luxurious 185-room Park Hyatt Seoul is a stylish 24-storey modern building fashioned like a glass box. The celebrated Japanese design firm, Super Potato, has made the idea of the hotel's exterior and interior one and the same, which makes for an impressive presence from street level and stunning panoramic views from within the hotel. Super Potato's concept melds Eastern and Western influences of tradition and innovation using nature as a guideline. Natural stone, light Myanmar oak, dark maple wood, water, and plenty of natural light are key elements throughout the hotel. The mood within the hotel changes constantly depending on the time of day, making for an endlessly surprising experience at every turn.

A spacious lobby on the top 24th floor and Park Club Spa, Swimming Pool and Fitness Centre also on the top floors showcase stunning city views. State-of-the-art rooms and suites with large bathrooms in sculptured granite with separate soaking bath and rain shower feature 3.4m floor-to-ceiling windows to enjoy city vistas.

The main restaurant Cornerstone with 360° open kitchen presents the most pleasant and romantic setting together with the sensory pleasures that stimulate all the five senses. The Lounge changes dramatically from day to night due to the abundance of natural light. While bright and comforting with a clear view of the busy streets and passersby in the bustling city, the night view captures the often missed side of city life creating the ideal mood for diners seeking to simply enjoy their meal while drinking in the view. With the design concept based on an old Korean wooden house, The Timber House providing a distinctive cultural space, displaying a gallery-like setting is divided into three thematic parts, with aspects of the house represented in each section: Sushi / sake / soju Bar, Cocktail Bar and Whisky Bar. In addition, Park Hyatt Seoul's hallways, lounge, restaurants, bars, and spa treatment rooms showcase about 300 sophisticated Korean art pieces and original antiques. Along with paintings and lithographs by famous Korean artists, traditional Korean Haheo masks, ceramics, old books and other antiques are displayed throughout the hotel for guests to fully experience the beauty of the Korean fine arts up close.

Completion date: April, 2005
Location: Seoul, South Korea
Size: 22,998m²

Designer: Super Potato
Photographer: Patrick Messina, Kyung Soo Kim,
Taeg Su Jeon

完成时间：2005 年 4 月
项目地点：韩国，首尔
面积：22,998 平方米

设计师：超级土豆
摄影师：帕特里克·梅西纳；金庆胜；全季洙

PARK HYATT

首尔柏悦酒店地处首尔金融、商业以及购物区江南区的心脏地带，共24层，拥有185间客房。酒店的建筑风格现代而时尚，宛如一个玻璃盒子。酒店的室内室外均由日本著名的超级土豆设计公司打造而成，外观气宇轩昂，室内视野开阔，可将城市景致尽收眼底。超级土豆的设计理念将东西方的设计风格融为一体，集传统与创新于一身，并将舒适自然作为首要的设计法则。天然石、缅甸橡木、深色枫木、水景以及大量的自然光线作为主要的设计元素，贯穿酒店的各个角落。酒店的氛围随着时间的变化而变化，总能给宾客带来不一样的惊喜。

酒店大堂位于24楼，空间宽敞开阔。嘉宾轩水疗中心、游泳池和健身中心分布在酒店的高楼层，俯瞰整个城市。客房和套房的设施一流，大型浴室采用雕刻花岗岩材料，设有独立的浸泡浴缸和淋浴间，高达3.4米的落地窗方便客人欣赏都市景致。

酒店的主餐厅Cornerstone拥有360度的开放式厨房，不仅氛围浪漫怡人，同时还为客人带来无尽的感官享受。由于不同时刻自然光线的不同，酒廊的氛围也会随之变化。白天的餐厅空间明亮舒适，视野清晰，将喧嚣的街道、行人以及繁华的城市一览无余。夜晚时分，都市久被遗忘的一面被重新拾起，餐厅氛围迷人，客人可以一边品尝美食，一边品饮绝妙夜景。

酒廊"The Timber House"的设计理念以韩国传统木屋设计为基础，为宾客提供一个别致的艺廊般的文化空间，整个空间被分为三个部分：寿司/清酒/烧酒吧，鸡尾酒吧以及威士忌酒吧。与此同时，首尔柏悦酒店还拥有大约300件精美的韩国传统艺术作品和古玩，分布在酒店的走廊、休息室、餐厅、酒吧以及水疗护理室等处。除了由著名的韩国艺术家创作的画作、石版画外，酒店的各个角落还陈设了一些韩国传统的Haheo面具、陶瓷制品、古书和其他古玩，方便客人全方位、近距离地感受韩国艺术的美丽。

1. Inspire and amaze guests with the sleek, modern design, stunning views, and five-star luxury of the premier wedding venue.
2. The Cellar, a luxurious private room with daylight in Cornerstone
3. Following the theme of the whole bar – Timber House, Whisky Bar is decked out with vintage wooden furniture.

1. 坐享美景，五星奢华的婚礼场地设计感时髦、现代
2. 基石餐厅的酒窖洒满了日光
3. 契合木屋餐厅的主题，威士忌吧装饰着复古木家具

1. The lounge absorbs the weather and natural illumination, changing dramatically by the minute.
2. With picturesque views of the city and dazzling sunlight streaming through the floor-to-ceiling windows, Cornerstone presents the most pleasant and romantic setting, perfect for special dining.
3. Infinity swimming pool located on the top 24th floor of the hotel, overlooking stunning city views through floor-to-ceiling windows.

1. 休息室天然采光照明，光线变化多姿
2. 光线透过落地窗洒落至室内，如画般的室外景色令基石餐厅展现出愉悦浪漫的完美就餐氛围
3. 在酒店顶层的超大游泳池，可以透过落地窗看尽城市美景

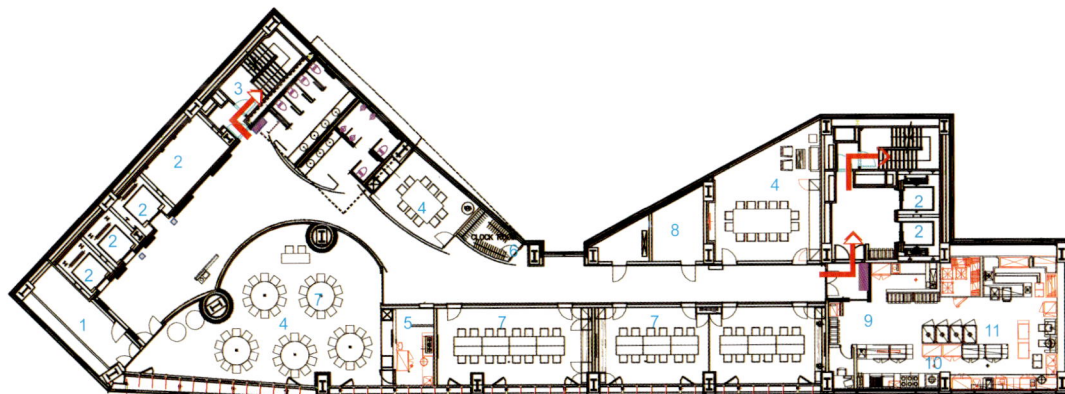

1. Banquet Storage	7. Board Room	1. 宴会储存室	7. 董事会议室
2. Lift	8. Alarm velve	2. 电梯	8. 报警阀
3. Fire Hydrant	9. Office Room	3. 消防安全栓	9. 办公室
4. Salon	10. Service	4. 沙龙	10. 服务
5. Pantry	11. Meeting Kitchen	5. 食品室	11. 厨房
6. Clock Room		6. 打卡室	

1. The unique gas fireplace in the living room of Presidential Suite reflects the three-dimensional ceiling.
2. Park Deluxe King is designed with natural elements of Myanmar oak and stone.
3. The dramatic bathroom in Presidential Suite
4. The bathroom in Park Deluxe King

1. 总统套房客厅里，独特的几何图案吊灯反映了天花板的三维效果
2. 柏悦豪华国王客房用天然材料——缅甸橡木、石头为设计用料
3. 总统套房夸张的浴室
4. 柏悦豪华国王客房的浴室

PARK HYATT BEIJING

北京柏悦酒店

Located in Beijing's CBD, Park Hyatt Beijing is in the central building of Beijing Yintai Centre on Changan Street, facing China International Trade Centre across the street. The hotel has 246 different rooms, including 25 suites. The guestroom's areas range from 45 square metres to 240 square metres. In addition, the hotel contains 2,864 square metres of conference facilities, including a 1,221-square-metre ballroom on the second floor, which could accommodate 1,200 guests. China Bar is on the 65th floor, while China Grill is on the 66th floor with high glass pyramid ceiling.

The 16 private rooms would provide traditional Cantonese cuisine and chef home cuisine on the 5th floor. Park Life Fitness Centre has glass ceiling, so the guests could enjoy the state-of-the-art fitness facilities in daylight. The fitness centre also contains a 25-metre indoor swimming pool, a fitness terrace for relaxation and training rooms for private and group training classes. Tian Spa on the same floor has 5 treatment rooms, providing various Spa programmes, combining Western-style Spa and Chinese elements perfectly.

Completion date: October, 2008
Location: Beijing, China
Area: 350,000m²

Designer: Remedios Siembieda Inc, Super Potato
& Bar Studio
Photographer: Michael Moran and Charlie Xia

完成时间：2008 年 10 月
项目地点：中国，北京
面积：350,000 平方米

设计师：雷梅迪奥斯·西艾姆别达设计
公司
摄影师：迈克尔·莫兰和查理·夏

地处北京 CBD 中央商务区核心地段的北京柏悦酒店坐落于长安街北京银泰中心中央主楼。内设包括套房在内的 246 间客房，风格别致，面积各异（45 平方米至 240 平方米）。同时设有 2,864 平方米的宴会会议空间。包括位于 3 层、占地 1,221 平方米的无柱大宴会厅，可容纳 1,200 位宾客。北京亮酒吧位于 65 层，配有高耸的玻璃金字塔天花的北京亮餐厅位于 66 层。主席台设有 16 间专属私人贵宾厅，提供广东传统粤菜及主厨私房菜。位于六层的悦•健身中心设有开放玻璃天花板，客人可以在健身的同时享受阳光。北京柏悦酒店拥有 25 米的室内游泳池、客人休憩的健身露台与可供私人及团体课程的训练室。在同一楼层的天池水疗中心，备有 5 间理疗室，提供多种水疗项目，将西式水疗与中式元素完美的结合在一起。

1. XIU – a rooftop garden bar features Chinese-style pavilions housing five themed bars.
2. One of 16 private dining suites is designed with its own living space.
3. One of the themed bars in XIU is furnished with Art Deco-style chairs.

1. 屋顶花园酒吧——秀吧采用中式楼阁设计，设有5个主题酒吧
2. 16间专属私人贵宾厅设有独立的起居空间
3. 一间主题酒吧采用了装饰艺术风格座椅

1. China Grill shares 360-degree views of Beijing with guests.
2. With the open kitchen, guests experience the most lively and electric cooking atmosphere.
3. Benefiting from the natural daylight and distinct lighting, the Salon offers flexible meeting space.

1. 北京亮餐厅为宾客提供了北京城360度全景
2. 开放式厨房让宾客可以体验最鲜活生动的烹饪氛围
3. 自然光和独特的照明将沙龙变为灵活的会议空间

1. A home library-inspired function room in Gallery, which is a residential event space.
2. The entrance of Gallery – the newest residential-style multifunction event venue is set with a stunning art piece.

1. 一间家庭图书馆风格的功能厅，可用于家居活动
2. 首创以豪宅府邸设计概念的"悦轩"，入口处装饰特色艺术品

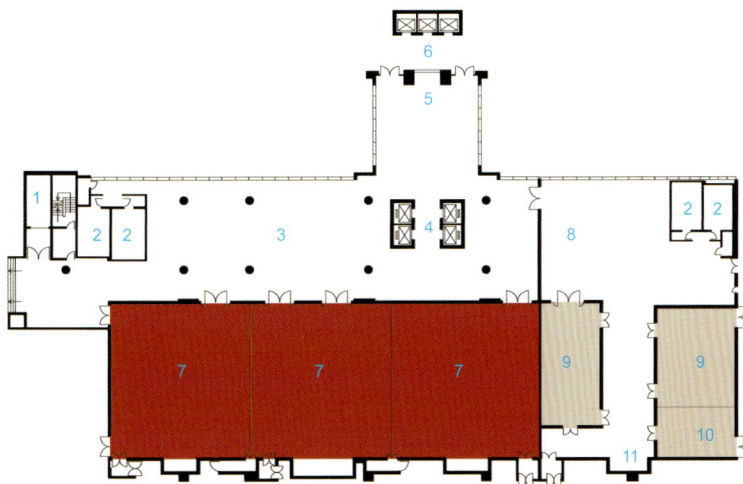

1. Car lift	1. 车用电梯
2. Restrooms	2. 洗手间
3. Prefunction	3. 迎宾区
4. Park Life lifts	4. 悦·生活电梯
5. Walkway to Park Life shopping	5. 通道至悦·生活购物
6. Hotel lifts	6. 酒店电梯
7. Ballrooms	7. 宴会厅
8. Courtyard	8. 前庭
9. Gallery	9. 悦轩
10. Library	10. 书房
11. Bar	11. 饮品区

1. Park King in contemporary décor and rich colours evokes a luxurious atmosphere.
2. Park Suite King has a spacious living room for resting and studying.
3. Penthouse Suite is decorated in modern minimalist style.
4. Park Deluxe King, an oversized, 50-sq.-metre corner room evokes the ambiance of a luxury residence with contemporary décor and rich woods, complementing a plush king-size bed and work desk.

1. 柏悦客房以其现代的装饰和丰富的色彩营造出奢华的氛围
2. 柏悦套房宽敞的客厅方便休息和工作
3. 顶楼套房采用了现代简约风格设计
4. 柏悦豪华客房,宽敞舒适的50平方米空间，简约实用的客房设计，拥有一张舒适的大床，宽敞工作台

GRAND HYATT GUANGZHOU

广州富力君悦大酒店

Located on one side of the beautiful Huacheng Square, Grand Hyatt Guangzhou is in the new axis of Guangzhou – Pearl River New City's CBD, which is a top-level business and social place in Pearl River Delta. The 350 rooms are decorated with fine and luxurious materials. The room innovatively divides the bath space into three areas: lavatory, dressing table and bathroom are separated. There are 5 restaurants and bars, all of which could enjoy a pleasant view of Huacheng Square. They're top preferences for high-end successful people to enjoy a private dining and a business dinner. Located in the sky bridge connecting two towers, Guanxi Lounge is a unique place. The semi-closed private sofa provides an intimate and comfortable space for meeting. The Grand Ballroom on the third floor is 979 square metres. With a fashionable and special diamond shape, the pillarless ballroom can be divided into three parts, thus being able to hold various pre-banquet cocktail events. The fitness centre and indoor swimming pool on the top floor will provide a panoramic view of Pearl River New City through their floor-to-ceiling windows.

Completion date: April, 2008
Location: Guangzhou, China
Area: 112,850m²

Designer: Remedios Siembieda, Super Potato,
Peter Remedios
Photographer: Samuel Nugroho

完成时间：2008 年 4 月
项目地点：中国，广州
面积：112,850 平方米

设计师：雷梅迪奥斯·西艾姆别达设计公司；
超级土豆公司；彼得·雷梅迪奥斯
摄影师：萨缪尔·纳格洛霍

1. Sky Lobby with its high irregular ceiling is separated by stunning timber wall.
2. Grand Club employs dark and steady decoration to create ideal backdrop for relaxing and rest.
3. With a casual yet modern ambience, The Market Café is a trend-setting buffet restaurant that offers Western and Asian cuisines from eight live cooking stations.

1. 空中大堂的不规则天花板被木板墙隔开
2. 嘉宾轩采用深色沉稳的装饰，为放松休闲提供了完美的背景
3. 凯菲厅是一家自助餐厅，8个现场料理台将在休闲而现代的氛围中提供中西美食

坐落于美丽的花城广场一侧，广州富力君悦大酒店位处广州新中轴线——珠江新城的商务中心区，交通非常便利，是珠江三角洲顶级的商务及社交场所。375间客房均由精美豪华的材质装设，房间创新地将传统的洗浴空间一分为三，洗手间、梳妆台和浴室各自独立。酒店共有五个餐厅与酒吧，均可欣赏花城广场的怡人景色，是高端成功人士的私人用餐及商务宴请首选。关系酒廊位于连接双塔的空中悬桥，是酒店最特别的地点之一，半封闭式设计的雅座给客人一个与密友惬意独享的空间。无柱式大宴会厅位于酒店三楼，面积979平方米，设计时尚独特，呈规则的梯形，并可灵活地分隔为两部分，能满足各种宴会前的鸡尾酒会等活动。位于酒店顶层的健身中心和室内恒温泳池，可透过落地窗俯瞰珠江新城的万千景色。

1. Stylish and contemporary Guangzhou restaurant – G Restaurant

2-3. The Grand Ballroom can accommodate up to 600 people for a banquet. The total area of 979 square metres can be used in its entirety or sectioned into three separate ballrooms.

1. 时尚且现代的广州特色餐厅—G餐厅

2、3. 大宴会厅总面积979平方米，可以容纳600人参加宴会；可以单独使用，也可划分为三个独立的小型宴会厅

1. The impressive sitting room in Chairman Suite holds dignified ambience.
2. Grand King features modern and minimalist design style. The stylish décor includes parquet and stone floors.
3. The elegant Bridal Room is decorated romantically with floating gauze.

1 总统套房的客厅散发出高贵的气息
2. 君悦客房采用现代简约设计风格；时尚的装饰包括镶木地板和石地板
3. 优雅的新娘房采用了浪漫的薄纱装饰

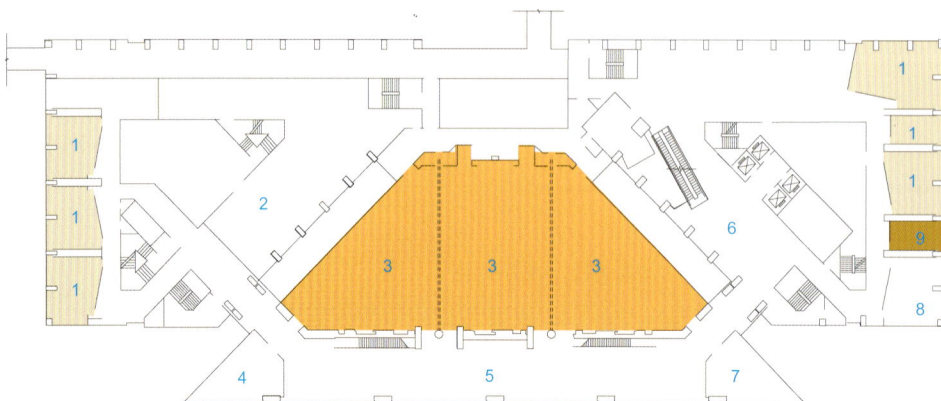

1. Salon	1. 沙龙
2. South tower	2. 南塔
3. Ballrooms	3. 宴会厅
4. Bridal room	4. 新婚房
5. Prefunction area	5. 迎宾区
6. North tower	6. 北塔
7. VIP room	7. 贵宾室
8. Business centre	8. 商务中心
9. Boardroom	9. 会议室

HYATT REGENCY HONG KONG, TSIM SHA TSUI

香港尖沙咀凱悅酒店

Hyatt Regency Hong Kong, Tsim Sha Tsui is located in Tsim Sha Tsui's business and tourism centre, which is a privileged location with convenient transportation. The hotel has 381 comfortable guestrooms and luxurious suites, with a gorgeous view of Victoria Harbour, Hong Kong Island or Kowloon. All rooms are equipped with state-of-the-art facilities, including LCD TVs, high-speed WiFi access and Ipod base, etc. The Regency Club on the 22nd to 24th floors accommodates guestrooms, lounge and meeting room. Through the floor-to-ceiling windows, you could enjoy a sweeping view of Victoria Harbour. The 592-square-metre meeting and private dining facilities are located on the lobby floor, including a pillarless ballroom and five meeting/event venues. The spacious and practical ballroom can be divided into two separate spaces according to requirement. The Chinese Restaurant combines the essence of modern art and 20s' teahouse, providing various Cantonese cuisine for the guests. In term of recreational facilities, there is an outdoor warm swimming pool on the 8th floor and a 24-hour gymnasium on the 10th floor, where the guests could get rid of their tires in the busy schedules.

Completion date: October, 2009
Location: Hong Kong, China
Area: 25,800m²

Designer: EKIT II DESIGN CO., LTD.; Elvis Kwan
Photographer: George Mitchell, Ong Yew Chuan

完成时间：2009 年 10 月
项目地点：中国，香港
面积：25,800 平方米

设计师：奥建建筑及室内设计有限公司；关志雄
摄影师：乔治·米切尔，王悠安

1. Hugo's is a fine dining restaurant offering authentic European cuisine. The décor in it is inspired by a fictional character.
2. With an eye-catching backdrop of stacked Chinese cupboards, Chin Chin Bar is a comfortable, cosy lounge bar with Chinese-style décor.

1. 希戈餐厅提供地道欧陆式美食；其室内装饰从一位小说人物中获得了灵感
2. 请请吧采用中式装饰，舒适而轻松，吧台后方摆放着引人注目的中式酒柜

香港尖沙咀凯悦酒店位处尖沙咀商业及旅游心脏地带，交通方便，占尽地利。酒店设有381间舒适客房及豪华套房，饱览维多利亚海港、香港岛或九龙的迷人景致（所有房间均备有先进设施包括平面液晶体电视、高速无线宽频上网系统及 Ipod 插座等）。位于酒店22至24楼的嘉宾轩设有客房、酒廊及会议室，透过落地玻璃窗将维多利亚港的迷人景色尽收眼底。会议及宴会设施位于酒店大堂楼层，占地592平方米，包括一个宴会大礼堂及五间宴会厅。无柱式宴会大礼堂宽敞实用，可按需要分为两间独立的宴会厅。位于3楼的凯悦轩餐厅糅合现代艺术及20世纪20年代传统茶馆的精粹，为客人提供各式地道粤菜佳肴。在康乐设施方面，室外恒温游泳池与24小时开放的健身室分别位于酒店8楼和10楼，让客人在繁忙的日程中尽洗疲累。

1. The Chinese Restaurant combines a modern Art Deco design masterfully blended with the essence of a traditional 1920s' teahouse, creating a warm and unique dining experience for authentic Chinese cuisine.

2. Part of Café is the library – a comfortable, residential-style space lined with books where diners can enjoy relaxing afternoon tea.

3. Diners may also host their privatised events or experience the exquisite Chef's Table in the 12-seat private dining room.

1. 凯悦轩将装饰艺术与20世纪20年代的茶馆巧妙地结合起来，打造出温馨独特的中餐环境

2. 咖啡厅的一部分被设计成图书阁，舒适的居家氛围适合享用悠闲的下午茶

3. 宾客可以在12人包房内举办私人活动或享用厨师晚宴

1. Hyatt Regency Hong Kong, Tsim Sha Tsui features a total of 592 square metres of meeting space on the lobby level, including a 335-square-metre ballroom with a five-metre ceiling, which has a perfect shape and can be easily partitioned into two meeting venues, as well as five meeting rooms of various sizes.

2. The luxurious private dining set-up in Regency Ballroom

3. All meeting facilities at the hotel are located on the same level and well-equipped with the latest technology and communications support.

1. 香港尖沙咀凯悦酒店拥有592平方米的会议宴会空间，其中包括335平方米的宴会大礼堂和5个大小不一的会议宴会厅；宴会大礼堂还可以被划分为两个独立的宴会厅

2. 宴会大礼堂豪华宴会布置

3. 酒店内所有的会议宴会设施都设在同一层，配有最先进的视听科技设施

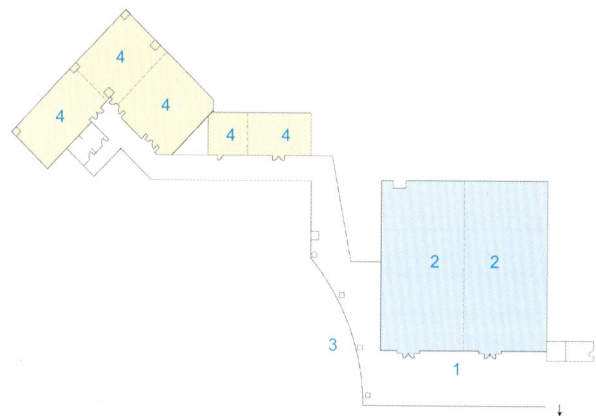

1. Pre-function area	1. 宴会大礼堂前厅
2. Regency Ballroom	2. 宴会大礼堂
3. Window	3. 窗
4. Salon	4. 宴会厅

1. Luxurious, top-floor 100-square-metre Executive Suite offers panoramic Victoria Harbour or Kowloon views, plus plush king bed, deluxe master bath with rain shower and separate tub, dining and living areas, guest bath and pantry area.
2. 150-square-metre Presidential Suite features cosy and modern décor in a plush living and dining area.
3. On the upper hotel floors, the Harbour View King room features panoramic views of Victoria Harbour, plus king bed in cosy and contemporary décor.
4. The bedroom in Presidential Suite features luxurious king bed.

1. 100平方米的行政套房享有维多利亚港或九龙的全景，设有舒适的大床、豪华的主浴室（配有淋浴和独立浴缸）、饭厅和客厅、客用浴室和茶水间
2. 150平方米的总统套房拥有舒适而现代的客厅和饭厅
3. 酒店上层的海景客房（特大床）享有维多利亚港的美景，配有舒适的大床和现代的装饰
4. 总统套房的卧室内设有豪华的大床

PARK HYATT SHANGHAI
上海柏悦酒店

Located in the eco skyscraper of Shanghai World Financial Centre in Lujiazui, Pudong, Park Hyatt Shanghai is one of the highest hotels in the world. The hotel is designed by award-winning New York designer Tony Chi. His design principle is to create a modern Chinese residence hotel, which will express a comfortable sense of a gorgeous house while respect Chinese traditional geometry and architecture. Entering from the entrance, guests will pass the lobby and guestrooms, finally to the courtyard. He imagines the courtyard as a peaceful and serene place, where guests could practice Tai Chi and play Chinese chess. Featuring beige colour and natural materials, the courtyard maintains exquisite and modest. Once you enter the hotels in a set of bamboos, all the hustle and bustle in Shanghai seems to fade away and a graceful experience of Park Hyatt Shanghai will start.

Completion date: September, 2008
Location: Shanghai, China
Area: 26,605m²

Designer: Tony Chi & Associates
Photographer: Michael Moran

完成时间：2008 年 9 月
项目地点：中国，上海
面积：26,605 平方米

设计师：季裕棠设计师事务所
摄影师：迈克尔·莫兰

1. The entrance to the lobby of Park Hyatt Shanghai
2. The entrance to the ground floor
3. The elegant counter of reception on 86th floor
4. Dining room features refined European elegant and serene setting.

1. 上海柏悦酒店大堂入口
2. 一楼入口
3. 87层优雅的前台
4. 餐厅以欧式优雅而宁静的布景为特色

上海柏悦酒店坐落于浦东陆家嘴金融贸易区中心区，环保型大厦上海环球金融中心内，是世界最高的酒店之一。酒店由屡获奖项的纽约设计师季裕棠先生设计。他的设计理念是设计出具有现代中国特色的私人住宅式酒店，在体现富丽堂皇之家应有的舒适之余，亦在设计中体现出设计师对中国传统几何学和建筑学的敬仰。宾客由大门开始，依次行经大厅、客房，最后以庭院为终点。他把庭院想象为一个平静、易于安神的圣地，客人可以在这宁静的庭院中练习太极和下中国象棋。庭院以自然米色为主色调，以天然物料为主要装饰材料，这样无论在感受或是设计方面均保持了中式的审美哲学理念——低调又不失讲究。客人一旦进入竹林掩映的酒店入口，上海的五光十色仿佛已经被抛于脑后，由此开始的将是在上海柏悦酒店的优雅体验。

1. Featuring a 12-seater dining table and a designer kitchen, the Chef's Table is ideal for tailor-made suppers and home-style hosting.
2. The intimate and exclusive leather-lines circular bar offers a rarefied ambience of exclusivity for guests.
3. Salons continue the philosophy of the Chinese private residence, including seven private meeting and dining rooms with daylight and striking city views.

1. 12人餐桌和设计厨房为家宴、晚宴打造了完美的气氛
2. 私密的圆形吧台为客人打造了考究的专属氛围
3. 沙龙延续了中式私人住宅理念，设有7间私人贵宾厅，均享有日光和城市美景

1. Salons
2. Entrance
3. Foyer
4. Lobby

1. 沙龙
2. 入口
3. 前厅
4. 大堂

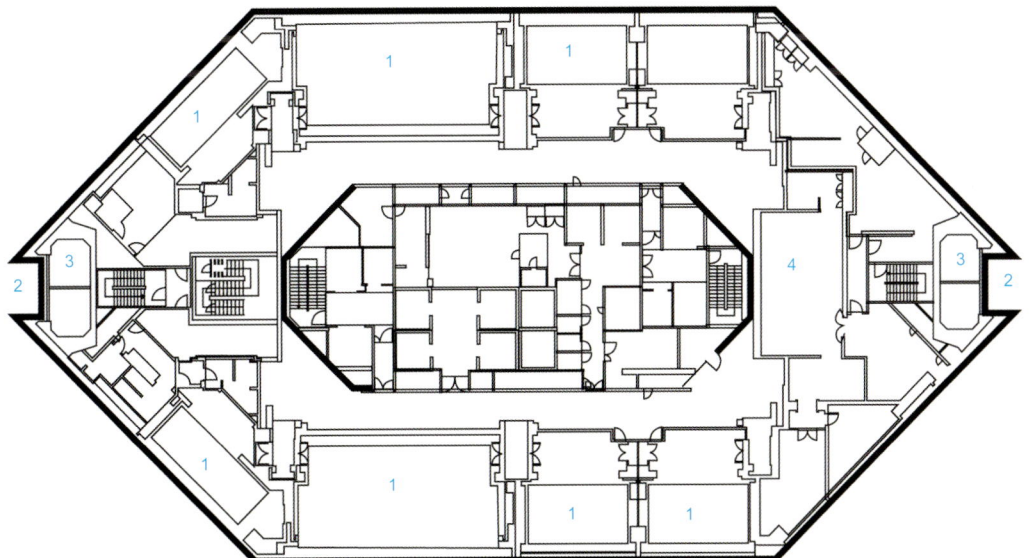

1. Water's Edge is dominated by a 20-metre infinity pool like the blue mirror.
2. Water's Edge is decorated in shimmering turquoise Mosaic-tiles, along with loungers with views of the city and the Huangpu River.
3. Bask in a luxurious, 194-square-metre residence – Chairman Suite that features separate living and dining rooms. The green carpet refreshes the elegant atmosphere.

1. 水疗中心20米长的水池宛如一面蓝色的镜子
2. 水疗中心装饰着闪闪发光的绿松石马赛克砖，享有城市和黄浦江的美景
3. 奢华的总统套房总面积194平方米，拥有独立客厅和餐厅；绿色的地毯活跃了优雅的氛围

1. With its wide space, Diplomatic Suite is designed in rich contemporary décor.
2. The luxurious and comfortable king beds in Chairman Suite
3. Park Suite King offers expansive city skyline or Huangpu River views, plus modern interiors that include one plush king bed.

1. 宽敞的外交套房采用了丰富的现代装饰
2. 总统套房内奢华舒适的大床
3. 柏悦套房享有上海或黄浦江的美景，配有舒适的大床等现代室内设施

HYATT REGENCY HANGZHOU
杭州凯悦酒店

Hyatt Regency Hangzhou on the shore of West Lake is a fashion landmark in Hangzhou. Located on Lakeshore Pedestrian street and embracing West Lake, the hotel has a privileged and convenient location, near the business, entertainment and shopping centre. The guestrooms are comfortable and elegant. Most of the rooms have stunning lake views through large observation windows, and are equipped with LCD TVs and state-of-the-art technology products (including WiFi access, doublet phones and voice mail boxes). The luxurious marble bathroom has an independent shower, a bathtub and a washing area. The 240-square-metre Presidential Suite located on the penthouse is the only suite with a 590-square-metre roof garden, an outdoor massage bath and a panoramic view of West Lake. There are also 11 elegant suites, including 8 Regency Suites, 2 Regency Executive Suites and a Garden Suite. The hotel provides a range of different restaurants and bars, each with a unique style and atmosphere. The lobby lounge creates a graceful and comfortable style, where guests could enjoy drinks or teas, coffee, desserts and snacks in an elegant atmosphere. The glass top lights of the indoor swimming pool will provide a beautiful view of West Lake.

Completion date: November, 2004
Location: Huangzhou, China
Area: 128,618m²

Designer: Florida Design Company, Heitz Parsons
Sadek, Jean-Philippe Heitz
Photographer: George Mitchell, Scott Wright

完成时间：2004 年 11 月
项目地点：中国，杭州
面积：128,618 平方米

设计师：美国佛罗里达设计公司；美国
霍普斯有限公司；让－菲利普·黑兹
摄影师：乔治·米切尔；斯科特·怀特

1. The palace-like lobby opens a journey of Chinese culture.
2. The 57-square-metre space of the Regency Club room is decorated simple yet graceful, providing a home-like atmosphere.
3. The marvellous 1,200-square-metre (12,916-square-foot) Ballroom can accommodate up to 860 people, and features a high ceiling, an excellent sound system and luxurious décor.

1. 宫殿般的大堂酒廊开启一段中国文化之旅
2. 57平方米的超大空间，嘉宾轩简单而高贵的室内装饰给人以舒适的居家之感
3. 华丽大气的宴会厅设计高雅，占地1,200平方米（12,916平方英尺），层高5米，可容纳860人的剧院式会议室，配备有先进的视听设备，是城中最大的宴会场地之一

1. Grand Banquet
2. Rest area
3. Greeting hall
4. Banquet hall

1. 悦华厅
2. 休息区
3. 迎宾厅
4. 大宴会厅

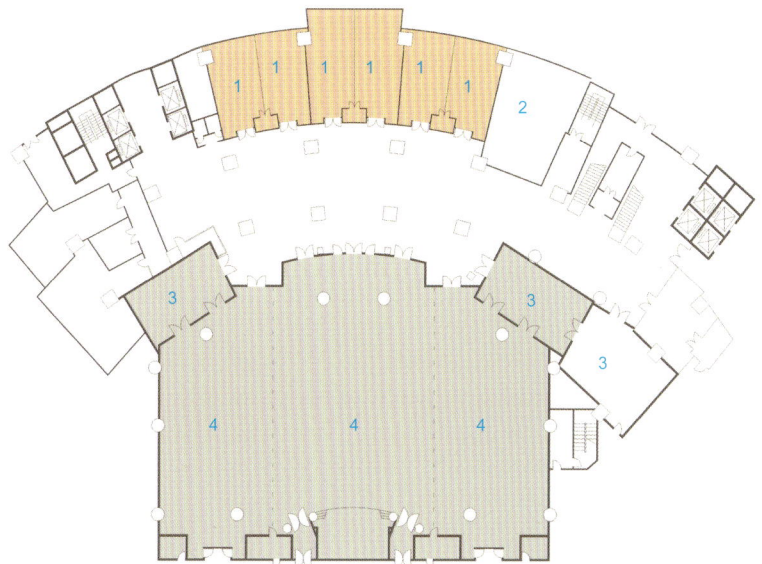

1

2

1-2. The various thematic banquet set-ups in meeting spaces
3. In the indoor swimming pool, several lively dolphin sculptures attract people's eyes, combining the interior environment and outside green landscape together.

1、2. 会议场所的各色主题宴会布置
3. 在酒店室内游泳池，几只生动的海豚雕像吸引人的眼球，将室内环境与窗外的绿色景观融为一体

坐落于西子湖畔的杭州凯悦酒店是杭州一个建筑性的时尚地标。酒店位于湖滨路步行街，坐拥风光如画的西湖，紧邻商务、娱乐及购物中心，交通非常便利。客房舒适典雅，多数房间拥有极好湖景、大幅观景窗、液晶电视，并配备最先进的客房科技产品，包括无线高速宽带网络接入、双线电话及语音留言信箱。豪华的云石浴室装设有独立花洒、浴缸及洁卫区域。酒店的总统套房，面积约 240 平方米，舒适得体，位于酒店的最高层，是城中唯一一间拥有 590 平方米屋顶花园的大套房、一个室外的按摩浴池以及全方位的西湖风光。酒店另有 11 间雅致套房，包括 8 间嘉宾轩套房、2 间嘉宾轩行政套房、1 间花园套房。酒店设置了多间不同类型的餐厅和酒吧供客人选择，每间餐厅均独具风格和气派。大堂酒廊营造出一种舒适典雅的风格，客人可以在优雅的气氛中享用美酒或茗茶、咖啡、甜点和小食。装有玻璃顶窗的室内温控泳池让您尽享西湖美景。

1. Presidential Suite provides a luxurious and elegant atmosphere with a reserved palette.
2. Garden Suite features a private garden connecting to the bedroom; guests could enjoy a magnificent landscape there.
3. The 57-square-metre space of the Regency Club room is decorated simple yet graceful, providing a home-like atmosphere.
4. The 36-square-metre Hyatt Lakeview Room overlooks West Lake. The modern artistic interior design is completed with various lighting facilities, providing an extraordinary lodging experience.
5. The bathroom in Regency Club Suite is equipped with shower and bathtub.

1. 总统套房整体以沉稳色调为基础打造奢华，优雅的氛围
2. 花园套房的特色是连通卧室的私人花园，华美景观亲自畅享
3. 57平方米的超大空间，嘉宾轩简单而高贵的室内装饰给人以舒适的居家之感
4. 从36平方米的房间饱览西湖全景，凯悦湖景房之名由此得来。室内现代化的艺术装饰品配以各色照明设施，给人非凡的入住体验
5. 嘉宾轩套房内，宽敞豪华的云石浴室配有独立花洒和浴缸

HYATT REGENCY
HONG KONG, SHA TIN

香港沙田凯悦酒店

Hyatt Regency Hong Kong, Sha Tin features 434 guestrooms and 133 long-stay rooms and suites, all commanding spectacular views of Tolo Harbour and Kau To Shan. There are five restaurants and bars: Café provides international and Hong Kong specialities; Sha Tin 18 Chinese restaurant offers home-style Dongguan cuisine and authentic Peking Ducks; Tin Tin Bar offers different kinds of Whisky and classic cocktails, with live musical performance; Pool Bar serves all-day drinks and snacks and weekend barbeque dinners; the special Sha Tin Apple Pie in Patisserie are fine gifts to take home. The hotel has spacious banquet venues, including a 30-table Ballroom, 3 Salons with floor-to-ceiling windows to introduce natural lighting, 9 mountain-view conference rooms on Regency Club, and a landscaped Garden for outdoor wedding and group training activities. Melo Spa offers a series of pomelo treatments in its 9 luxurious suites and Melo Moments relaxation suite which can sit up to 12 guests. Other recreational facilities include a fitness centre, a tennis court, Camp Hyatt for children, a 25-metre-long outdoor swimming pool, whirlpool, sauna and steam bathrooms.

Completion date: Feburary, 2009
Location: Hong Kong, China
Area: 80,651m^2

Designer: Steve Leung
Photographer: George Mitchell, Ong Yew Chuan

完成时间：2009 年 2 月
项目地点：中国，香港
面积：80, 651 平方米

设计师：梁志天
摄影师：George Mitchell, Ong Yew Chuan

1. The reception area is the focus in the design of the lobby.
2. Regency Club provides comfortable sofas and pleasant talking atmosphere.
3. Tin Tin Bar's warm tone and modern interior design provides a privileged yet relaxing atmosphere.

1. 独具艺术气息的接待区是酒店大堂的设计重点
2. 嘉宾轩提供舒适的沙发与交谈氛围
3. 在天天吧，温暖的色调与现代的室内设计提供给客人优越且放松的氛围

香港沙田凯悦酒店位于港铁大学站旁，备有 434 间客房及 133 间服务式长租客房及套房，均可眺望吐露港、九肚山或沙田马场景致。另外，酒店设有 5 间餐厅及酒吧：咖啡厅提供国际及香港地道美食；沙田 18 供应家庭式东莞菜及正宗北京烤鸭；天天吧供应多款威士忌及经典鸡尾酒，亦设现场音乐演奏；池畔酒吧于假日提供烧烤晚餐；饼店的特色沙田苹果派则是送礼佳品。酒店的宴会场地包括可容纳 30 席的宴会大礼堂、3 间设有大型落地玻璃引入天然日光的凯悦厅、9 间位于嘉宾轩楼层的山景会议室，及适合举办户外婚礼及团体训练活动的园林草地。Melo Spa 水疗中心设有 9 间豪华套房及设 12 个座位的"欢聚时光"护理室。在此，客人可接受一系列特色柚子护理疗程，其他休闲设施包括健身中心、网球场、专为儿童而设的凯悦儿童营、25 米室外恒温泳池、按摩浴池、桑拿及蒸汽浴室。

1. There are 4 show kitchens with orderly layout in the Café.
2. Sha Tin 18 dessert kitchen features Chinese elements, including wood window lattices and custom lantern pendants.

1. 咖啡厅设有4个开放式主题厨房，场景布置非常规整
2. 沙田18甜点厨房设计特色以中式元素为主，包括木质的窗棂，设计感十足的仿灯笼吊灯

1. The Regency Club Executive Suite on the top floors is 120 square metres, enjoying a gorgeous view of Tolo Harbour.
2. The 133 specially designed serviced suites for extended stay provide residential comfort in a stylish hotel setting.
3. Deluxe room facing mountains measures 67 square metres, with comfortable queen size bed, a large desk, extra closet space, and a dining area.

1. 嘉宾轩行政套房面积达120平方米，套房位于酒店高层，坐拥吐露港醉人景致
2. 在酒店时尚的布局之中，133间特别设计的套房为客人提供如家般舒适的居住环境
3. 奢华套房面朝群山，约67平方米，内有舒适的双人床，大写字台，独立衣橱及就餐空间

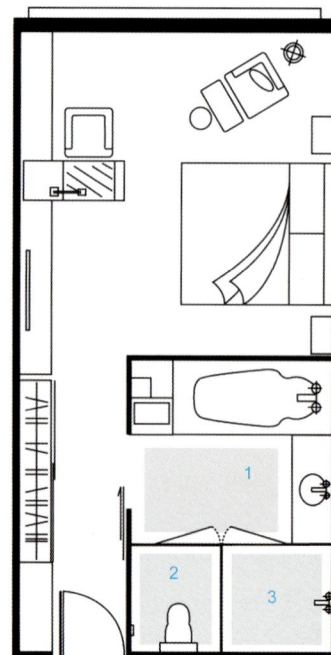

1. Bathroom
2. Washroom
3. Shower

1. 浴室
2. 洗手间
3. 淋浴室

MANDARIN ORIENTAL HOTEL GROUP
The Oriental Elegance and Luxury

文华东方酒店集团——来自东方的淡雅奢华

MANDARIN ORIENTAL
THE HOTEL GROUP

Founded in 1963, this young hotel group enjoys a world-wide reputation for its oriental accents and traditions. The name "Mandarin Oriental" comes from the first two hotels of the group: Mandarin Hotel Hong Kong and Oriental Hotel Bangkok. One follows Chinese traditional design in Westernisation, and the other highlights oriental reserved luxury in Thai culture. Afterwards, the brand characteristics of these two hotels are blended into more than 40 Mandarin Oriental hotels' architectural and interior designs: oriental elegance and moderate luxury.

With the lead of the two flagship hotels, the design of Mandarin Oriental hotels achieves a precise balance – a reserved oriental feature in occident form. This feature runs through every Mandarin Oriental hotel's architectural and interior design. Take Mandarin Oriental Las Vegas for example. Borrowing Chinese traditional patterns and colours in the façade design, KPF linked aluminium and glass vertical panels together. The indirect use of oriental elements also enhanced the sustainability, since the interlaced patterns on the façade form transparency in various levels, introducing daylight into the deeper space of the hotel. In addition, more Mandarin Oriental hotels use oriental elements in the interior design, although there might be quite a cultural difference in some places. In Mandarin Oriental San Francisco, the interior design extensively uses Chinese elements: the window lattice is round, which is regularly seen in Chinese gardens. In the luxurious guestroom, old Chinese furniture echoes with the hanging crystal chandelier. Around the European style sofa is the tea table with carved patterns and some rattan chairs, which looks nothing but luxurious. In Mandarin Oriental Munich, a hotel with Neoclassicism, you can also find several smooth figures of Buddha or some small Chinese stone lions.

The logo of Mandarin Oriental is an unfolded fan with 11 folds. A folding fan reflects the elegance of Chinese culture and is also the embodiment of status in ancient Chinese culture. In Mandarin Oriental's culture, the folding fan represents the reserved elegance and luxurious beauty of the hotel while the hotel design is a direct expression of this oriental beauty. In the architectural and interior designs of Mandarin Oriental, you can rarely see expensive art works nor furniture made of rare materials. The hotel will express a luxurious atmosphere through its harmonious overall design. In Mandarin Oriental Singapore, the guestroom employs textured ivory wallpapers to highlight the overall elegant atmosphere through this pure colours. The polished natural wood bedhead highlights harmony through natural material.

In a Mandarin Oriental hotel, guests can feel not only the abundant and thoughtful board and lodging reception, they can also sense the unique oriental accents with elegance and luxury. Without visiting every Mandarin Oriental hotel in person, you could experience the oriental elegance and luxury through the four well-selected hotels in this chapter.

这家年轻的酒店集团成立于 1963 年，像它的名字一样，以它的东方风情与传统，享全球美誉。"文华东方"这个名字来源于集团最早的两家酒店：香港"文华酒店"以及曼谷"东方酒店"。这两家酒店，一个在城市西方化中遵循中式传统设计，一个则在泰式文化中凸显东方的含蓄奢华。而此后，这两家酒店所体现出的品牌特质也融汇在全球成立的 40 余家文华东方酒店的建筑与室内设计之中：淡雅东方，中庸奢华。

在两家旗舰酒店的带领之下，文华东方酒店的设计有了一种精准的平衡感——在西方的形式之下蕴含着不张扬的东方特色。这一点贯穿至每一个文华东方酒店的建筑与室内设计之中，例如拉斯维加斯文华东方酒店，KPF 事务所在对酒店建筑立面进行设计时借鉴了中国传统的图案和色彩，将铝制和玻璃垂直板材相互连接在一起，这种在建筑设计中间接采用东方元素的手法还起到了提升酒店可持续性的作用，因为立面上相互交织的图案形成各种层次的透明感，可以将日光引进酒店更深的空间当中。此外更多的文华东方酒店在室内设计方面采用东方元素，即使在有相当文化差异的地区，例如旧金山文华东方酒店，其室内设计将中国元素大量运用，窗棂设计成中式园林中常见的圆形，与室内悬挂的水晶吊灯相映成辉，在奢华的客房内还可见老式的中国家具。雕花的茶几，藤椅摆设在欧式沙发周围，增设奢华气息。又如文华东方慕尼黑酒店，在新古典的风格基调下，竟然也会在大厅、客房，甚至浴室内发现几个圆润的佛像，或是中国小石狮。

文华东方酒店的 LOGO 为十一折展开的折扇。折扇正反映了中国文化中的优雅之美，也是中国古代文化中身份地位的体现。折扇在文华东方的酒店文化中也正代表着酒店不张扬的优雅，奢华之美，而酒店的设计又正是这种东方中庸之美的直接呈现。在文华东方酒店的建筑与室内设计之中，很少看见昂贵的艺术品，或奇珍材料制成的家具，而是通过酒店各部分和谐的整体设计效果来体现酒店的奢华。例如文华东方新加坡酒店，在酒店的客房，贴着带纹理的象牙白墙纸，采用这种素雅的颜色烘托整体气氛的优雅，并在床后装饰上抛光的天然木板，用自然材料突出和谐之感。

在文华东方酒店，客人感受到的不仅仅是丰富且周到的食宿接待，细心徜徉，还有东方韵味绕梁不绝，淡雅奢华随处相伴。在本章，不用亲身奔赴各文华东方酒店，精选的 4 家最具代表性的酒店将带您细数酒店的东方淡雅奢华。

MANDARIN ORIENTAL, TOKYO

东京文华东方酒店

Creating a distinctive "Sense of place" is a core guiding principle of the Mandarin Oriental, Tokyo, located in Nihonbashi, the centre of the city formerly known as Edo. While retaining its connection to Edo's past, the ultra-modern Mandarin Oriental, Tokyo embodies the respect and appreciation for the natural world that the Japanese people have possessed since time immemorial.

Though it would undoubtedly surprise many visitors to this densely urbanised country, Japan is a land of forests. The nation's 25 million hectares of woodland covers 67% of its land area, making it one of the most forested countries in the world. Within a country surrounded by water, Japan's mountains give birth to clouds that nourish its soil with abundant rainfall, and feed the many rivers that flow from the mountains to the coastlines. Therefore, "Woods and Water" became inspirational themes and formed the underlying design construct of Mandarin Oriental, Tokyo. Fabrics and furnishings have also been created to seamlessly weave together Nihonbashi's past, present, and future. Starting with an insistence on discerningly selected natural materials, Japanese designer Reiko Sudo relied upon both her own considerable talents, and those of countless master artisans and weavers throughout the country, to produce original fabrics that express the artistic and cultural traditions of Japan in colour, pattern, and texture.

The hotel has been conceived as a single large living tree, providing shelter, comfort, and a gathering place for the community. These themes are expressed using imaginative, custom fabrics and furnishings throughout, including wallpaper, carpets, upholstery, drapery, cushion covers, etc. The spiritual elements of the beauty of individual things, and the importance of understatement and reserve, convey traditional Japanese aesthetic values. No single object imposes itself on the visitor's senses. Rather, all of the adornments blend harmoniously, eliciting a response through subtle, restrained representation.

Completion date: 2005
Location: Tokyo, Japan
Designer: LTW Designworks Pte Ltd (Lobby, K'shiki,
Gourmet Shop, Oriental Lounge, Chapel, Spa,
Meeting rooms, Ballroom, Guestrooms
& Suites)
Photographer: George Apostolidis
Area: 100,000m²

完成时间：2005 年
项目地点：日本，东京
设计师：LTW 设计公司（大堂，四季亚洲餐厅，美食店，
东方酒廊，礼拜堂，Spa，会议室，宴会厅，客房和套房）
摄影师：乔治·阿普斯托里迪斯
面积：100,000 平方米

1. The east lobby offers the spectacular view via floor-to-ceiling windows, while the Japanese style chandeliers light the gracious interior scenery.
2. The verdant bamboos in Ventaglio – the Italian restaurant reflects the design theme of the hotel – "Woods and Water".

1. 东侧大堂的落地窗展示了外面壮观的美景，而日式吊灯则让室内空间更显优雅
2. 温塔格里奥意大利餐厅里青翠的竹子反映了酒店的设计主题——"树木和水"

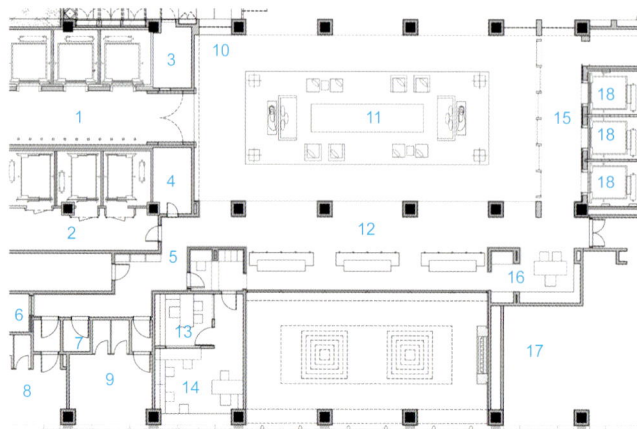

English	中文
1. Lift hall 3	1. 电梯大厅3
2. Service lift	2. 服务电梯
3. PS	3. 机电室
4. PS/EPS	4. 机电室/蓄电室
5. Public phone	5. 公用电话
6. H/Capped toilet	6. 残疾人洗手间
7. Janitor	7. 警卫室
8. Ladies	8. 女洗手间
9. Gents	9. 男洗手间
10. Shutter	10. 防火门
11. Hotel main lobby	11. 酒店大堂
12. Reception	12. 前台
13. Office	13. 办公室
14. Front office	14. 前台办公室
15. Hotel lift lobby	15. 酒店电梯大厅
16. Concierge	16. 门房
17. GYM	17. 健身房
18. Lift	18. 电梯

位于日本桥地区的东京文华东方酒店的设计以独一无二的"空间感"为指导方针。这一地区从前以江户而闻名。超现代的东京文华东方酒店保持了与江户时代的联系，充分尊重并欣赏了日本人从古时一直流传下来的自然世界。

尽管许多游客会感到惊讶，但是作为一个如此高度密集城市化的国家，日本拥有大面积的森林。25,000,000公顷的林地占据了全国面积的67%，是全球森林面积最大的国家之一。作为四面环水的国家，日本的山脉高耸入云，而云以丰富的降雨滋润了土壤，并且供养了许多从高山流向海岸的河流。因此，"树木和水"成为了设计的灵感主题，形成了东京文华东方酒店的基本设计结构。织物和家具装饰与日本桥地区的过去、现在和未来天衣无缝地交织在一起。日本设计师须藤礼子坚持选择自然材料，她凭借自己出色的才华和无数的工匠、纺织工一起，在色彩、图案和材质上体现了日本传统工艺和文化。

酒店被设计成一棵巨大的树，为人群提供避难所、舒适和集会的场所。这些主题通过形象的定制织物和装饰得到了体现，其中包括墙纸、地毯、装饰物、帷帐、套垫等。独立物件所呈现的精神之美以及低调和朴素的重要性反映了日本传统美学价值。任何一件物品都不会凌驾于客人的感觉之上。所有的装饰品都和谐地混合在一起，通过精致、克制的表现激发人的回应。

1. The luxurious carved screens and artistic fabrics make a heart-stopping impression once the guests enter Signature – the French-inspired restaurant.
2. In Asian-inspired restaurant – K'shiki, the understated colour palette offers an ideal venue for relaxation with fantastic views towards the lively business district of Tokyo.

1. 奢华的雕刻屏风和艺术织物在署名法式餐厅为宾客留下了深刻的印象
2. 色彩朴素的四季亚洲餐厅是休闲的好去处，能够远眺东京繁忙的街景

1. On the 37th floor located off the main lobby Oriental Lounge is the hotel nucleus.

2. In the design of Sense, the Cantonese restaurant, two elements – water and fire interplay to create a dramatic atmosphere.

3. A microcosm of futuristic luxury – Mandarin Bar, like a gallery of design, incorporates an astounding blend of surfaces, textures and shapes.

1. 位于38楼的文华东方酒廊是酒店的核心

2. 感觉广东餐厅中，水与火两个元素相互交织，形成了戏剧化的氛围

3. 文华酒吧是未来奢华的缩影，宛如设计的画廊，结合了震撼人心的平面、纹理和造型设计

1. Reaching through the atrium to the ceiling above is a wall of fire; this fabulous structure is the visual focus of the 36th floor Sense Tea Corner.
2. Chapel in the hotel offers a sense of purity for celebrating special time.
3. Linden is the largest banquet venue within the Mitsui Main Building.

1. 从中庭直达天花板的是一面火之墙；这个非凡的结构是37楼Sense茶吧的视觉焦点
2. 酒店内的小教堂渗透出一种纯净感，适合庆祝特殊的时刻
3. 林登厅是三井主楼内最大的宴会场所

1. In the living room of Presidential Suite, the unique "Isegata", a pattern for Yukata, mounted on the wall is a collector's.
2. The spacious Dynasty Suites preserve a distinct sense of Japanese aesthetics in the selected fabrics and fine art.
3. The Executive Suites offer sweeping views. The rare "Isegata" hung on the wall tells the Japanese culture of hospitality.

1. 总统套房的客厅墙上采用了独特的日式浴衣图案，具有收藏价值
2. 宽敞的王朝套房展示了独特的日式美学，采用了精选的织物和艺术品进行装饰
3. 行政套房享有辽阔的视野；墙上珍贵的浴衣布料展示了日式文化的精髓

1. With natural textiles and fallen leaves in wallpaper, Deluxe Room is designed to reflect Japanese reverence for nature.

2. Ideal suites for business arrangements, the living and entertaining area is carefully separated from the private quarters and thoughtfully designed to accommodate meetings.

3. Equipped with luxurious spa bathroom amenities, Mandarin Suite Bathroom is a model of luxury, an irresistible invitation to recline in the spa tub with a view of the busy metropolis over 30 floors below.

4. In the Oriental Suite Bathroom, relaxing in the marble bathtub makes a sumptuous leisure base.

5. In the king bedded rooms the designer tub is right by the glass, luxuriates and enjoys the view.

1. 墙纸上的自然纹理和落叶图案让豪华房的设计反映了日本人对自然的崇敬

2. 商务客人的理想套房，起居室与娱乐区与私人空间小心地分隔开来，方便安排会面

3. 文华套房浴室内配有奢华的卫浴设施，是奢华的典范；水疗浴缸吸引着人们躺下并欣赏外面都市的风景

4. 东方套房浴室的大理石浴缸尽显奢华

5. 大床房内的设计浴缸紧靠窗户，可以享受窗外的景色

MANDARIN ORIENTAL, SINGAPORE

新加坡文华东方酒店

Mandarin Oriental, Singapore is located in Marina Centre adjacent to Marina Square Shopping Mall, within walking distance of Suntec City, which houses one of Asia Pacific's largest convention centres, the Suntec Singapore International Convention and Exhibition Centre, and Marina Bay Sands Resort and Casino.

The harmonious décor in the elegant guestrooms takes its inspiration from the natural world, and the colour palette is neutral and soothing. Guests have the choice of a king-size bed or two double beds. Natural elements can be seen in the polished wood panel behind the bed; the bed spread with a motif that resembles twigs, and the textured ivory cloth on the walls. Pan-Asian influences are apparent in the choice of decorative details, with a nod to Singapore's multi-cultural heritage, and the overall effect is one of subtlety with a decidedly contemporary twist.

Sleek granite and dark timber bathrooms are spacious, with both a bathtub and a separate glass-enclosed shower. Attractive timbered blinds on the floor-to-ceiling glass panel that separates the bathroom from the living space, can be left closed for privacy or raised to reveal the breathtaking view from the bedroom window.

The Oriental Spa in Mandarin Oriental Singapore is located on level 5 of Mandarin Oriental, Singapore. The 528-square-metre spa comprises four luxurious treatment rooms, a Couples' Suite, a Shiatsu room, a private Relaxation Lounge, a mind and body yoga patio and a high-performance fitness studio.

The innovative and restorative treatments at The Oriental Spa merge both ancient and modern techniques and philosophies from around the world, using 100% pure essential oils and herbs. The Oriental Spa aspires to provide each guest with a personalised sensory journey to wellness, focusing on the six senses. With its exotic oriental interiors, coupled with the authentic spa rituals from skilled therapists, The Oriental Spa presents its guests with a complete luxury spa indulgence experience.

In keeping with its sense of place and reinforcing the exotic roots of Mandarin Oriental, The Oriental Spa is both contemporary and reflective of the surrounding Asian cultures. Tranquility and harmony are the inspiration behind the design of the spa; features include walnut timber flooring, Asian motif panels, complemented with traditional Chinese furniture.

Completion date: 2005
Location: Singapore
Designer: LTW Designworks Pte Ltd (Lobby, Melt, Oriental Club, Cherry Garden, Axis Bar & Lounge, Dolce Vita, Meeting Rooms, Ballroom, Gym,

Guestrooms & Suites)
Photographer: Mandarin Oriental Hotel Group & George Apostolidis
Area: 14,500m²

完成时间：2005 年
项目地点：新加坡
设计师：LTW 设计公司（大堂，MELT，东方酒廊，樱桃园，Axis 酒吧＋酒廊，Dolce Vita, 会议室，宴会厅，健身房，

客房及套房）
摄影师：文华东方酒店集团和乔治·阿普斯托里迪斯
面积：14,500 平方米

1. The soft lightings in the lobby shine on the timbered backdrop, causing elegant atmosphere.
2. With panoramic views of the Singapore skyline, the Oriental Club is carefully designed for offering a peaceful retreat.

1. 大堂内柔和的灯光照在木背景上，营造了优雅的氛围
2. 精心设计的东方俱乐部提供了平和的休息场所，享有新加坡天际线的美景

新加坡文华东方酒店位于滨海中心，紧邻滨海购物广场，距离新达城仅有几步之遥。新达城内拥有亚太地区最大的会议中心、新达新加坡国际会展中心和滨海湾金沙酒店及赌场。

新加坡文华东方酒店客房

优雅的客房中那些融洽的装饰的设计灵感来源于自然世界，而色彩搭配则显得中性而舒缓。客人们可以选择一张特大号床或两张双人床。床后的抛光木板呈现出自然元素；床单和墙上的象牙色织物上带有树枝图案。装饰细节的选择呈现出显著的亚洲色彩，同时还增添了新加坡的多文化特征。整体效果精致而现代。

由光滑的花岗岩和深色木材装饰的浴室十分宽敞，带有浴缸和独立淋浴间。落地玻璃板上独具魅力的木制百叶窗将浴室与起居空间隔开，既可以闭合起来保证隐私，又可以打开来看到卧室窗外的美景。

新加坡文华东方酒店东方水疗中心位于酒店五层的水疗中心占地528平方米，由四间奢华的理疗室、一间情侣套房、一间日式指压室、一间私人休息室、一个身心瑜伽露台和一间高效健身工作室组成。

东方水疗中心的创新和滋补治疗源于世界各地的古老和现代技术与理论的结合，采用100%纯度的精油和草药。东方水疗中心为每位客人提供了个性化感官健康之旅。具有东方异国情调的室内设计与治疗师所提供正宗的水疗程序共同为东方水疗中心的客人提供了全套的奢华水疗体验。

为了保证自身的空间感和文华东方的异国情调，东方水疗中心兼具现代特色和亚洲文化风格。平静与和谐是设计的灵感来源；水疗中心的特色设计包括胡桃木地板、亚洲图案墙板和传统中式家具。

1. Stepping through Cherry Garden, the traditional courtyard style makes everyone stun by the modern design of oriental splendor.

2. Up to 600 guests can enjoy an uninterrupted view of the proceedings – amazingly there are no pillars.

1. 穿过樱桃园，传统的庭院风格让人们被现代设计的东方光彩所叹服

2. 600名宾客可以享受大堂一览无余的景色，里面没有一根柱子

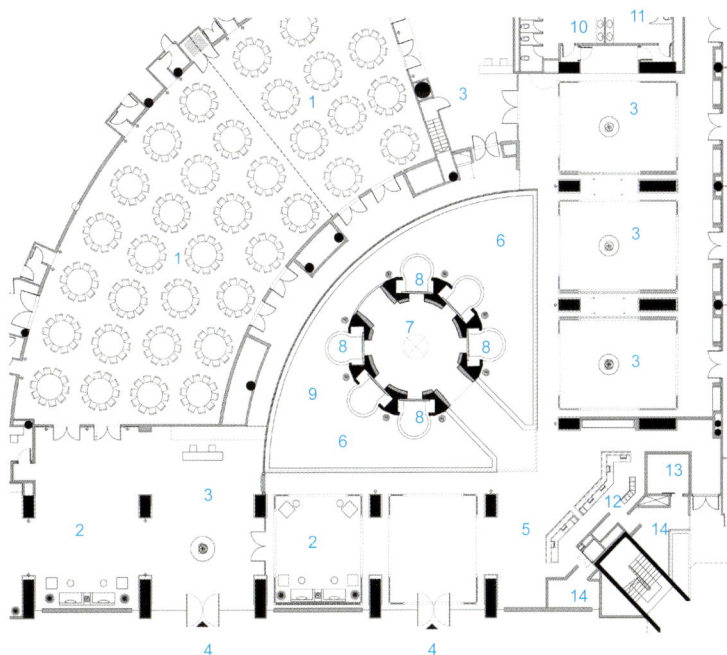

1. Ballroom   1. 宴会厅
2. Waiting area   2. 等候区
3. Pre-function   3. 准备区
4. Main entry   4. 主入口
5. Main lobby   5. 主大堂
6. Water body   6. 水体
7. Lift lobby   7. 电梯大厅
8. Lift   8. 电梯
9. Artwork   9. 艺术品
10. Ladies   10. 女洗手间
11. Gents   11. 男洗手间
12. Reception   12. 前台
13. Front office   13. 前台办公室
14. Office   14. 办公室

1. Luxury in every sense, the Presidential Suite combines fine contemporary style with delicate oriental touches.

2. In the Bay Suite Bedroom, natural elements can be seen in the polished wood panel behind the bed, the bed spread with a motif that resembles twigs, and the textured ivory cloth on the walls.

3. Premier Harbour rooms are designed with restrained luxury, reflecting oriental touches in silks, colours and details. With a flick of the Venetian timber blinds guests can enjoy the superb harbour view from the bathroom.

4. The Premier Suite is one of our grandest suites, with one bedroom and a great deal of space to entertain in style. The sleek open spaces are designed with subtle oriental touches.

5. Mandarin Grand is designed with a generous seating area and larger working space. They comfortably accommodate a rollaway bed making these a popular choice for families. Earth tones, soft silks and natural elements all create a calm retreat for travellers to this exciting city.

1. 总统套房在各种意义上都彰显了奢华，结合了现代风格和精致的东方气息

2. 海湾套房卧室的自然元素体现在抛光木制床头板、床单上的树枝图案以及墙壁上的象牙色织纹布上

3. 高级海港景观房具有内敛奢华，折射出丝绸、色彩和细节的东方风格。拉开威尼斯木质百叶窗，客人可以从浴室欣赏到壮丽的海港景色

4. 高级套房是酒店最豪华的一间套房，配备有一间卧室和宽敞的娱乐空间。开放式的空间融入了古老的东方风情

5. 文华至尊房配备有宽敞的座位区和较大的工作空间。房间舒适地容纳一张折叠床，是家庭的最佳选择。泥土色调、柔软的丝绸和天然元素为到这座充满活力的城市旅游的旅行者创造了一个避世胜地

MANDARIN ORIENTAL, BOSTON

波士顿文华东方酒店

Taking cues from the days of the Silk Trade, when ships regularly plied the ocean between Boston and the Orient, award-winning designer Frank Nicholson elegantly fuses the grace and refinement of ancient Asian design with the modern sensibilities of cosmopolitan Boston in the interior design for Mandarin Oriental, Boston.

In its design, each new Mandarin Oriental hotel aims to infuse nuances of Asia with the essence of the city in which it resides. The design model for Mandarin Oriental, Boston is to combine subtle Asian elements, with a sense of place reflecting the ambiance and identity of Boston. The building's heavy stone limestone and marble façade is designed with an architectural theme reminiscent of Boston's distinctive brownstones. Inside the hotel, exotic blonde wood panelling reflects Boston's historic style, complemented by a modern motif of rich textures and fabrics that evoke the elegance of the Orient.

A residential, comfortable feel is key to the design of the guestrooms. Using elegant materials in a contemporary Asian design, each room offers seating for two guests, custom-made furnishings including a club chair, full-size sofa, and writing desk. Rest is essential to those seeking an urban resort in Boston. Each room is outfitted with opulent features and luxurious silks in warm, buttery colours or cool jade green tones. All guestrooms feature Ploh duvets and pillows, luxurious 400-thread Frette linens, and a 42" high definition flat screen TV. Each four-fixture bathroom features a soaking tub, rain shower head, water closet, and marble sink. Luxurious Frette bathrobes, Aromatherapy Associates bath products and a large vanity area complete this peaceful space. The overall atmosphere of the guestrooms has been personalised for the hotel, weaving classic Bostonian elements with a touch of Asian elegance.

Two striking features of the Forbes Four-Star rated signature restaurant Asana are a limestone wall with a distinctive hand-carved texture and warm, rich bamboo flooring designed to give the room a standout presence. The tabletops are set with custom granite in ebony and Inca gold colours. A distinctive, specially designed table seating eight guests with a rift-cut ebonised white oak base and a custom etched mirror table top is illuminated from underneath to create a distinctive and unique dining experience, featuring a seasonal menu of authentic and pure New England cuisine.

Completion date: 2008
Location: Boston, USA
Interior Designer: Frank Nicholson Incorporated
Architect: CBT Architects

Photographer: Mandarin Oriental Hotel
Group & George Apostolidis
Area: 929m²

完成时间：2008 年
项目地点：美国，波士顿
室内设计师：弗兰克·尼科尔森公司

建筑师：CBT 建筑事务所
摄影师：文华东方酒店集团和乔治·阿普斯托里迪斯
面积：929 平方米

丝绸贸易时代，波士顿和东方之间定期有航船往来，获奖设计师弗兰克·尼科尔森从中获得了灵感，将古亚洲设计的优雅和精致与波士顿的现代风格结合在一起，融入了波士顿文华东方的设计之中。

波士顿文华东方酒店的风格

在设计中，每家新建的文华东方酒店都在其所在的城市特色中融入了亚洲风格。波士顿文华东方酒店的设计模型结合了微妙的亚洲元素和波士顿典型特征。建筑厚重的石灰岩和大理石外立面令人回忆起波士顿独特的褐色砂石建筑。酒店内部，具有异国风情的亚麻色木护墙板反映了波士顿的历史风格，而丰富的材质和纹理所形成的现代图案则唤起了东方的优雅。

客房和套房

舒适的居家感是客房设计的关键。现代亚洲设计风格之中融入了优雅的材质，每间客房都为两位客人提供了休息空间。定制的家具包括扶手椅、沙发和书桌。放松休息对于在波士顿寻求都市度假村的人们来说至关重要。每间客房都配有丰富的特质，奢华的丝绸采用了温暖的黄油色调或者清冷的翠绿色。所有客房都配有普罗牌被子和枕头、奢华的400罗纹弗雷特床单和42寸高清平板电视。每间浴室都配有浴缸、淋浴花洒、洗手间和大理石水槽。奢华的弗雷特浴袍、香薰协会洗浴用品和宽敞的化妆区进一步完善了这个宁静的空间。客房的整体氛围被酒店个性化，交织了经典的波士顿元素和亚洲风情。

阿萨纳餐厅

由福布斯评选的四星级餐厅阿萨纳餐厅拥有两个显著的特色：具有独特雕刻纹理的石灰岩墙壁和丰富的柱子地板，二者为空间带来了杰出的表现。定制的大理石桌面上镶有乌木和印加金色。独一无二定制八人桌由斜径刨切的乌木化白橡木底座和定制的蚀刻玻璃桌面组成。桌子被从下方照亮，营造出独特的就餐氛围。餐厅以地道纯粹的新英格兰四季美食为特色。

1. Inside the hotel, exotic blonde wood panelling reflects Boston's historic style.
2. The lounge in the lobby is complemented by a modern motif of rich textures and fabrics that evoke the elegance of the Orient.

1. 酒店内具有异域风情的亚麻色木镶板反映了波士顿的历史风格
2. 大堂的酒廊采用了纹理丰富的现代图案装饰，显示了文华东方酒店的优雅

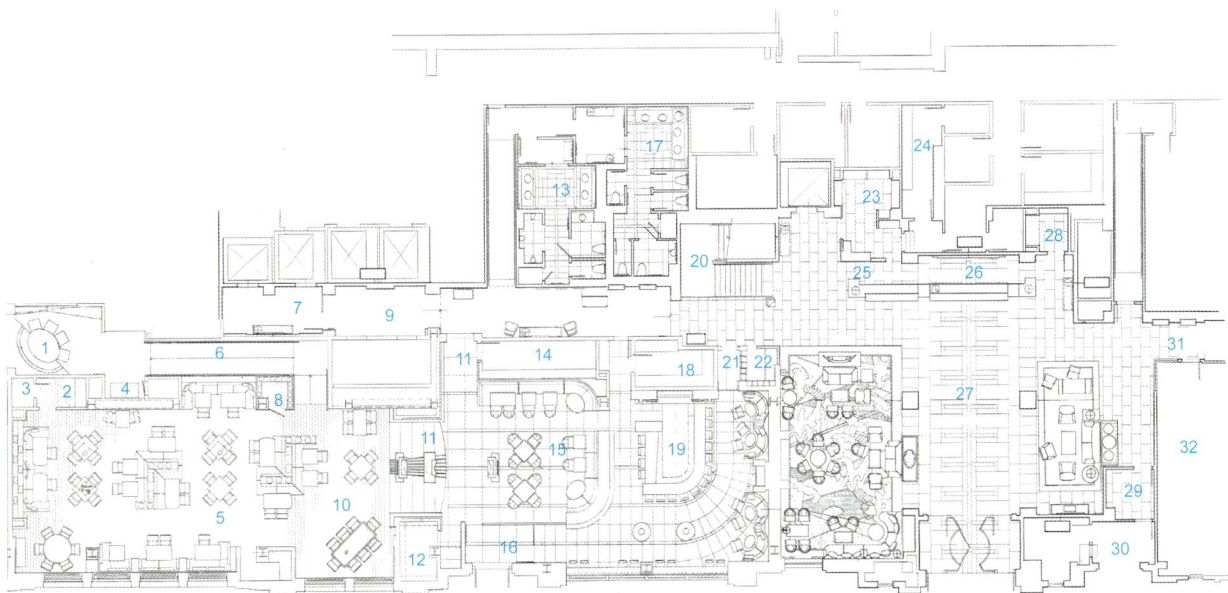

1. Private dining	12. Coats	23. Sundries	1. 包房
2. Service	13. Men	24. Concierge storage	2. 服务区
3. Dining accessories closet	14. Wine room	25. Concierge	3. 餐具橱柜
4. Display	15. Lounge	26. Front desk	4. 展示区
5. Restaurant	16. Entry vestibule	27. Lobby	5. 餐厅
6. Ramp	17. Women	28. S.D.V	6. 坡道
7. Condominium lift lobby	18. Pantry	29. Baggage	7. 公寓电梯大厅
8. Cheese room	19. Bar	30. Valet	8. 奶酪室
9. Guest lift lobby	20. Grand stair	31. To residence lobby	9. 客用电梯大厅
10. Wood floor	21. Bar foyer	32. Retail	10. 木地面
11. Stone floor	22. Tea service		11. 石地面

12. 衣帽间	23. 杂货间
13. 男洗手间	24. 门房储藏室
14. 酒廊	25. 门房
15. 休息室	26. 前台
16. 入口门廊	27. 大堂
17. 女洗手间	28. 机械室
18. 备餐间	29. 行李寄存处
19. 酒吧	30. 服务台
20. 大楼梯	31. 通往住宅大堂
21. 酒吧门厅	32. 商店
22. 茶水服务	

1. The tabletops in Asana are set with custom granite in ebony and Inca gold colours.
2. Beijing Room – one of additional function rooms, has natural light with views over Boylston Street.

1. 阿萨纳的桌面采用了黑色花岗岩和印加金色装饰
2. 北京厅是一间额外的功能厅，光线充足并享有波伊斯顿街的景色

1. Asana is a limestone wall with a distinctive hand-carved texture and warm, rich bamboo flooring designed to give the room a standout presence.
2. The chef's table in Asana with modern décor, where the cooking process can be seen.
3. Crystal Chandeliers matching charming flowers, Oriental Ballroom offers a happiness ambience.

1. 阿萨纳是一面带有独特手刻纹理的墙壁，而温暖丰富的竹地板为房间营造了出色的效果
2. 阿萨纳餐厅的餐桌采用了现代装饰，可以边就餐边观看烹饪过程
3. 水晶吊灯与迷人的花朵相得益彰，东方宴会厅提供了欢快的气氛

1. Dynasty Suite Living Room's colour scheme accents dark-toned furniture with champagne and sterling blue drapery creating a space that is elegant and stylish.

2. Champagne-coloured drapery adorns the windows of the large master bedroom and the finest Frette linens, Ploh pillows and duvet dress the king-sized bed. Oriental Suite is an oasis of elegance and comfort that features a refined palette.

3. The plush master bedroom in Dynasty Suite, with its silk gold-coloured wall coverings, is furnished with a king-sized bed dressed in the finest materials and a custom-designed white maple hardwood headboard.

4. Premier Suite is outfitted with a full-size sofa and custom-made chair in warm colour. The décor is inspired by the refined elegance of traditional New England.

5. With oversized bathrooms and opulent décor, the Mandarin Rooms provide the perfect urban escape.

1. 王朝套房客厅的色彩以深色调家具为主，配有香槟色和蓝色的帷幔，打造了优雅、时尚的氛围

2. 香槟色的窗帘装饰着主卧室的窗口，而精致的弗雷蒂床单、保罗枕头和羽绒被则点缀了大床。东方套房是优雅和舒适的绿洲，以精致的色彩搭配为特色

3. 王朝套房豪华的主卧室采用了丝绸金色壁纸装饰；特大号双人床上铺有精致的床品和定制的白枫木床头

4. 尊贵套房配有暖色的全套沙发和定制座椅；室内装饰从新英格兰传统风格中的精致优雅中获得了灵感

5. 特大的浴室和丰富的装饰让文华房在城市中打造了完美的世外桃源

THE LANDMARK MANDARIN ORIENTAL, HONG KONG

香港置地文华东方酒店

The stunning new Landmark Mandarin Oriental is one of Asia's most luxurious, stylish and contemporary hotels, with some of the biggest hotel rooms in Hong Kong and also one of its premier holistic spas offering a comprehensive range of Eastern and Western treatments.

The hotel is located at The Landmark in the heart of Central, a few steps from most of Hong Kong's key commercial buildings and shops, the area providing the highest concentration of luxury brands in Asia. This premier location, combined with international architect-designed restaurants and bars serving superlative food and wine, ensures the hotel will be an irresistible magnet for Hong Kong's social and business elite. In addition, purpose-built function rooms are perfect for designer fashion shows, product launches and other celebrations.

Anthony Costa, General Manager, says, "The Landmark Mandarin Oriental has established a new benchmark for luxury, elegance and dedicated service combined with state-of-the-art technology and the latest in contemporary design."

It is the first time that world-renowned hotel interior designers, Hong Kong-born Peter Remedios and Transylvanian Adam D. Tihany, have worked together. Remedios, who is based in Los Angeles, designed the 113 stunning rooms and suites and the spa, while Tihany created unique restaurant and bar concepts destined to be Central's new icons.

The hotel offers the biggest hotel rooms in Hong Kong, with an average size of more than 50 square metres – indeed most rooms are more than 600 square feet – a true indulgence in a busy metropolis where ample space to relax is a luxury. Every aspect is designed for pure sensory indulgence, from the avant-garde objects d'art and the stunning glass-walled bathrooms, to the 400 thread-count linen and the revolutionary in-room guest entertainment, with most rooms having three LCD televisions.

Completion date: 2009
Location: Hong Kong, China
Designer: Remedios Studio

Photographer: Mandarin Oriental Hotel Group & George
Apostolidis
Area: 15,000m²

完成时间：2009 年
项目地点：中国，香港
设计师：雷梅迪奥斯工作室

摄影师：文华东方集团和乔治·阿普斯托里迪斯
面积：15,000 平方米

1

1. Hotel lobby guides you to a new luxury and elegance journey in contemporary design.

2. The provocative ceiling sculpture is truly a piece of art, presenting cutting-edge design of Amber.

3. For welcoming the guests, the lobby lounge is designed in rich warm palette.

1. 酒店大堂引领着客人进入一个奢华优雅的现代设计之旅

2. 夸张的天花板雕塑是一件艺术品，呈现了先锋设计元素

3. 大堂休息室采用了温暖的色调，让人有宾至如归的感觉

置地文华东方酒店是全亚洲最奢华的现代风格酒店之一，拥有香港最大的酒店客房，而它的顶级综合水疗中心将提供全套的东西方疗法。

酒店位于中环置地广场，距离香港的主要商业大厦和购物中心仅有几步之遥，当地汇聚了亚洲最顶级的奢华品牌。绝佳的地理位置与国际知名建筑师设计的餐厅酒吧所提供的美食美酒让酒店吸引了大量香港社交界和商界精英。此外，酒店的功能厅还是举办设计时装秀、产品推介会和其他庆典活动的完美场所。

酒店的总经理安东尼·科斯塔称："置地文华东方酒店创建了奢华酒店的新标准，它结合了奢华、优雅、贴心的服务、最先进的技术以及最新的现代设计。"

项目是世界知名的酒店室内设计师彼得·雷梅迪奥斯（香港）和亚当·D·蒂哈尼（特兰西瓦尼亚）的首次合作。雷梅迪奥斯负责设计113间绝佳的客房和套房以及水疗中心，而蒂哈尼所设计的独一无二的餐厅和酒吧注定会成为中环的新地标。

酒店拥有香港最大的客房，平均面积超过50平方米——大多数房间都超过了55平方米，在繁华的都市中提供了奢侈的大型休息空间。酒店的每个方面都为纯粹的感官纵情体验而设计，从前卫的艺术品和绝妙的玻璃墙面浴室到400罗纹床单和革命性的客房娱乐设施（大多数客房都配有液晶电视），一应俱全。

1. L450 Deluxe is granted with sophisticated design and supremely comfortable.
2. With imported Italian marble and an astute selection of contemporary art, L900 Landmark Suite offers the most sophisticated experience.
3. L600 Deluxe features a luxurious 19-square-metre circular glass bathroom.
4. Featuring an open plan lounge and bedroom, L600 Executive offers a sense of free-flowing space.

1. L450豪华房拥有精致的设计和顶级的舒适
2. 进口的意大利大理石和精心挑选的现代艺术品让L900置地套房为客人打造最高级的体验
3. L600豪华房以19平方米的圆形玻璃浴室为特色
4. L600行政房拥有开放式布局的休息室和卧室，是个流畅的空间

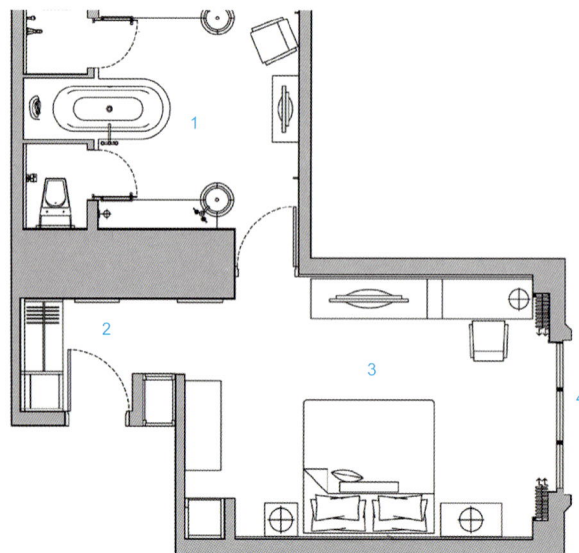

1. Bathroom
2. Vestibule
3. Bedroom
4. City view

1. 浴室
2. 门廊
3. 卧室
4. 城市景观

1

1. Set with a spectacular circular bathroom, L600 Premier resembles a private spa – a blend of contemporary design and cutting-edge technology.

2. L450 Superior owns a marble bathroom.

3. With clean lines and subtle colours, the spectacular open plan bathroom in L900 Landmark Suite is full of contemporary art.

1. 华丽的圆形浴室让L600顶级房好像一个私人水疗馆，房间结合了现代设计与先锋技术

2. L450高级房拥有一个大理石浴室

3. 简洁的线条和精妙的色彩让L900置地套房的开放式浴室充满了现代艺术气息

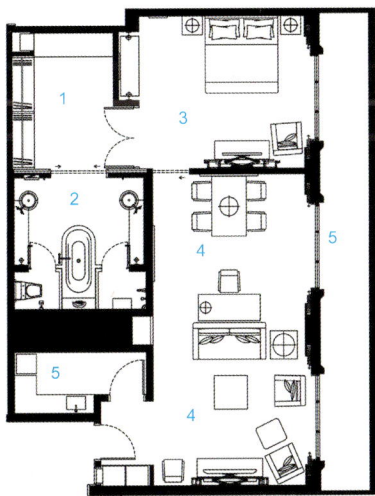
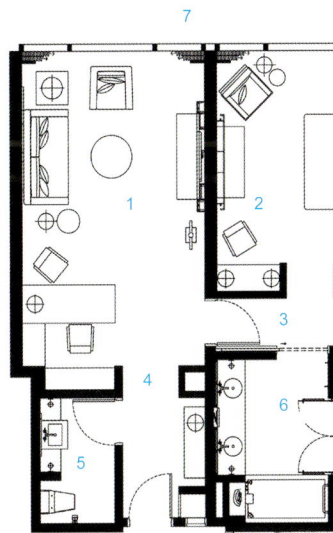

1. Walk-in closet
2. Bathroom
3. Pantry
4. Living room
5. Balcony
6. Statue square view

1. 步入式衣柜
2. 浴室
3. 备餐间
4. 客厅
5. 阳台
6. 雕像广场景观

1. Living room
2. Bedroom
3. Walk-in closet
4. Vestibule
5. Powder room
6. Bathroom
7. Habour view

1. 客厅
2. 卧室
3. 步入式衣柜
4. 门廊
5. 化妆间
6. 浴室
7. 海景

SHANGRI-LA HOTELS AND RESORTS
An Earthly Paradise

香格里拉酒店集团——现世的世外桃源

SHANGRI-LA
HOTELS and RESORTS

The Shangri-La story began in 1971 with their first deluxe hotel in Singapore. Today, with 72 hotels and resorts throughout Asia Pacific, North America, the Middle East, and Europe, the Shangri-La group has a room inventory of over 30,000. Shangri-La Hotels and Resorts is Asia Pacific's leading luxury hotel group and also regarded as one of the world's finest hotel ownership and management companies.

Shangri-La is an earthly paradise in the Himalaya, mysterious and peaceful, noble yet extraordinary. It is composed of eight areas with lotus petal shapes, with a beautiful snow mountain in the centre. There is no diseases or quarrels in Shangri-La, but only non-fading blossoms, running water and sweet fruits. People live in this paradise harmoniously and peacefully. Inspired by the legendary land featured in James Hilton's 1933 novel, Lost Horizon, the name Shangri-La encapsulates the serenity and service for which Shangri-La Hotels and Resorts is renowned worldwide. The perfect and peaceful environment of Shangri-La hotels rightfully explains this mysterious name.

Shangri-La Hotels and Resorts transforms the legendary paradise into reality in their architectural and interior designs. The logo S refers to the unique form of Asian architecture. The upper part of S represents a mountain; the lower part represents the reflection; the line in the middle distinguishes the two. The whole logo looks like a magnificent mountain reflected in a peaceful lake. The logo design reflects the typical eave and outline of an Asian architecture, which is also the representation of the hotel's architectural spirit. Shangri-La hotels reflect Asian architecture's eternal elegance, monumental serenity and extraordinary comfort.

The design of Shangri-La hotel features fresh landscape design, excellent and elegant lighting design and deep Asian cultural flavours. The hotel is normally located in a graceful natural landscape, which follows the tradition of "scene" in Asian architecture. Shangri-La Hotel Beijing has lavish landscape design, including five landscape areas. The tea lounge garden is decorated with flowerpots, paved with unique paving and planted with evergreen flora. In the lobby spring area, flowers and water emphasise the floating water lilies. The Chinese pavilion landscape area features well-proportioned plants. The wood pavilion landscape area is specially designed for wedding celebration. The leisure landscape area provides relaxing space with a pavilion along the creek. In addition, Shangri-La hotels always use Asian decorations to create an elegant and luxurious atmosphere. Each Shangri-La will feature some Chinese landscape paintings and calligraphies. Another feature is the lighting design. The crystal chandelier is a brand DNA in Shangri-La hotels. The subtle and excellent lighting designs add graceful atmosphere for the luxury. Shangri-La Hotel Tokyo has 50 crystal chandeliers, 20 of which are custom designed. The grand chandelier on the ground floor is inspired from the natural beauty of ginkgo leaves. Standing with it, as if you could hear the whispers of the leaves in autumn sunshine. The splendid carved chandelier expresses a natural grace instantly.

This chapter selects eight hotels from Shangri-La Hotels and Resorts, including seven Shangri-La hotels and a Kerry hotel. The gorgeous scenes of the hotels will present you an earthy paradise.

1971 年，香格里拉酒店集团在新加坡设立的第一家豪华酒店。时至今日，香格里拉酒店集团的 72 家酒店及度假酒店遍布亚太地区、北美、中东和欧洲，拥有 30,000 多间客房。酒店集团迅速成长为亚太地区最豪华的酒店集团，同时是世界上公认的最佳酒店管理公司之一。

"香格里拉"这个名字原是传说中喜马拉雅的一个人间天堂，神秘而幽静，高贵而脱俗。它由八个成莲花瓣状的区域组成，中央耸立着美丽的雪山。在这里没有疾病与争吵，只有花常开水常绿，甜蜜的果子常挂枝头，人们和谐安宁的生活在这片乐土。1933 年，英国作家詹姆斯·希尔顿发表了小说《消失的地平线》。书中详述了香格里拉———一个安躺于西藏群山峻岭间的仙境，让栖身其中的人，感受到前所未有的安宁。时至今日，香格里拉已成为世外桃源的代名词，它所寓意的恬静、祥和、殷勤的服务，完美地诠释了闻名遐迩的香格里拉酒店集团的精髓，而香格里拉酒店完美憩静的环境，正与这个弥漫着神秘色彩的名字源出一辙。

怎么样把传说中的世外桃源转化为现实，香格里拉在酒店的建筑与室内设计上做到了这一点。就像酒店的标识"S"，标识借鉴了亚洲建筑的独特形式，S 上半部分代表一座山峰，下半部代表水中倒影，中间连接线代表区分二者的连接线，犹如雄伟壮丽的山脉倒映在宁静的湖面上。标识的设计反映了典型的亚洲式建筑的屋檐顶和曲线构思，也正是酒店建筑精神的体现。亚洲建筑永恒的优雅，不朽的祥和，非凡的舒适都是香格里拉酒店的设计精髓。

香格里拉的设计一向以清新的景观设计、高超典雅的照明设计，以及富有浓厚亚洲文化气息特征闻名于世。酒店通常被置于优美的自然景观之中，这一定程度上沿承了亚洲建筑重视"景"的传统。例如北京香格里拉酒店，酒店有着丰富的景观设计，景观区包括五个部分，有装饰花钵，铺置特色铺地，种植四季常绿植物的茶座景观区；有飞花渐水，睡莲漂浮的大堂涌泉水景区；有植物错落有致的中式亭廊景观区；还有一婚庆为主题的木亭景观区以及溪流伴寄，为客人提供休闲空间的景观区。此外，香格里拉酒店常用亚洲文化特色装饰出高雅的奢华氛围。每一家香格里拉酒店都会使用中国的山水画，或者笔墨珍品。香格里拉酒店的另一个设计特色，是酒店的照明设计。其中堪称香格里拉酒店"品牌 DNA"的是每家酒店必备的水晶吊灯，每一家酒店都因各自巧妙而高超的照明设计将酒店的奢华蒙上一层典雅的气息。例如东京香格里拉，酒店内的水晶吊灯共有 50 余盏，其中的 20 余盏是为酒店特别定制。位于酒店一楼的巨大水晶灯从银杏叶的自然之美中获取灵感，仿佛能聆听到秋日阳光里叶子在微风轻拂下发出的沙沙声，充满雕琢痕迹的华丽水晶灯瞬时绽放出一种天然的优美。

本章精选 8 家香格里拉集团酒店，包括 7 家香格里拉酒店和一家嘉里酒店，一步一景，现世的世外桃源为您跃然纸上。

SHANGRI-LA HOTEL, GUANGZHOU

广州香格里拉大酒店

Strategically located in the city's new business and commercial district, the Shangri-La Hotel, Guangzhou is adjacent to the Guangzhou International Convention and Exhibition Centre— home to the world-renowned Canton Fair and the largest exhibition centre in Asia, with easy access to a host of major attractions and nearby transportation.

Nestled amidst 5,800sqm of tranquil gardens overlooking the Pearl River, the hotel is a reputed urban oasis, offering 704 spacious rooms and suites and 26 serviced apartments. Eight stylish restaurants and bars cultivate an authentic ambience for exploring dynamic culinary arts and a newly renovated backyard putting green allows for a friendly golf practice session of "approach and putt". 6,000sqm of meeting and banqueting venues, featuring two majestic ballrooms, eight function rooms and a 100-seat auditorium, cater to events of any scale. In addition, the signature CHI, The Spa at Shangri-La blends traditional Asian healing philosophies with modern-day luxury to create an experience of pure indulgence and spiritual revitalisation.

Completion date: 2008
Location: Guangzhou, China
Designer: HBA

Photographer: Shangri-la Hotels and Resorts
Area: 43,000m²

完成时间：2008 年
项目地点：中国，广州
设计师：HBA

摄影师：香格里拉酒店集团
面积：43,000 平方米

1. Guangzhou Ballroom foyer offers both refined interior layout and splendid exterior views.
2. coolThai restaurant offers a stylish new entertainment venue.
3. Private room at il Forno Italian restaurant is designed like a wine cellar vault, one side with brick walls and authentic leather, and the other side linked to a temperature-controlled wine cellar, in order to create a cosy and private ambience.

1. 广州宴会厅前厅既拥有典雅华丽的室内布局,又享有广阔的江景视野
2. 香泰泰国餐厅提供了时尚的休闲就餐场所
3. 爱弗罗意大利餐厅贵宾房一面墙身以红砖和意大利真皮砌成、另一面则与酒库相连,别具匠心

广州香格里拉大酒店位于城中新商业区的中心地带,傍依风光秀丽的珠江之滨,毗邻亚洲最大的、举办各种大型会议及每年两届广交会的广州国际会展中心。交通条件优越便捷,轻松可抵城中各大旅游景点。

酒店坐拥5,800平方米的优雅花园,是城中的一片绿洲。704间豪华宽敞的客房以及26套配套完善的服务公寓,为每一次居停缔造舒心的休憩体验。八间风格迥异的餐厅及酒廊汇集世界各地美馔;新建成的高尔夫球场尽显高尔夫短距离果岭推杆及沙地切杆练习的无穷乐趣。面积达6,000平方米的会议及宴会场地包含两个气派非凡的大宴会厅、八间多功能会议厅以及可容纳100名宾客的礼堂,为各种规模的宴会提供宽敞灵活的空间。此外,香格里拉特有的"气"Spa让宾客静享私人空间,暂别急促的都市节奏,以畅快的感官之旅唤醒生命能量,让身心重新绽放神采。

1. Terrace	5. The auditorium	1. 平台	5. 礼堂
2. Xi Qiao	6. Stage	2. 西樵	6. 舞台
3. Gui Feng	7. Tian Lu	3. 圭峰	7. 天露
4. Lian Hua	8. Ding Hu	4. 莲花	8. 鼎湖

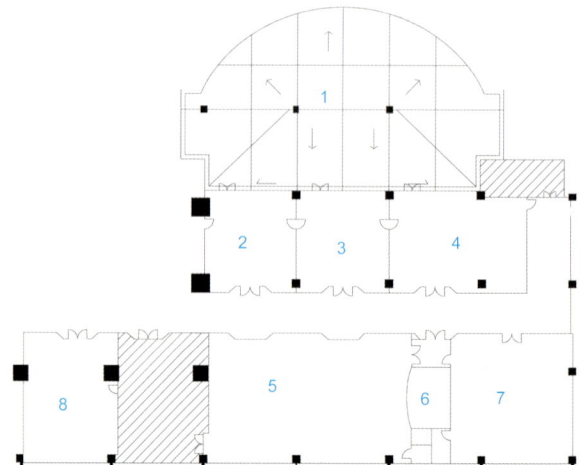

1. Summer Palace is inspired by Chinese imperial banquet, offering royal luxury.
2. The lounge at Horizon Club features delicate art pieces on the wall and various crystal pendant lights hanging under the European ceiling.
3. Compared with other space, Lift Bar has a more lively atmosphere.
4. With three teppayaki private rooms, an open teppanyaki counter and a sushi counter, diners get a splendid view of the gardens, the city and the Pearl River whilst enjoying the most authentic Japanese dining experience at Nadaman.

1. 夏宫从中国皇室宴会中获得了灵感，尊贵而奢华
2. 豪华阁休闲室内，墙上装饰着精致的艺术品，欧式的吊顶下悬挂着各式水晶吊灯
3. 与其他空间相比，交点吧拥有更活泼的氛围
4. 滩万日本料理内设有三间铁板烧包房、一个开放式铁板烧吧台和一个寿司吧，就餐者可以边享受花园、城市和珠江的美景，边享用正宗的日本美食

1. The pillarless Guangzhou Ballroom with high ceiling has the capacity to accommodate any kind of banquets.
2. Equipped with a grand built-in stage and the latest of audio visuals, Pearl River Grand Ballroom will make any event an exceptional experience.
3. Tian Lu Function Room is ideal for the VIP reception.

1. 广州宴会厅采用无柱式挑高天花板设计，可举办各种宴会
2. 巨大的嵌入式舞台和最新的视听设备让珠江厅为各色活动打造了非凡的体验
3. 天露厅是高级别贵宾接待的理想场所

1. Shangri-La's oriental elegancy spreads throughout the décor of living room at Horizon Deluxe Suite.
2. With exquisite Chinese tea table, the pure white colour palette makes Shangri-La Suite unique luxury.
3. Horizon Premier Suite is designed with a dining area for private gathering in the living room.
4. Upholstered sofa, king bed with silk covers, beautiful embroidery, all of these echoes the theme of oriental elegancy.
5. The Horizon Club Riverview Room has floor-to-ceiling windows, as well as a large work area for your comfort.

1. 豪华阁套房的客厅装饰尽显香格里拉的东方优雅
2. 古朴的中式茶桌和纯白的色调让香格里拉套房显得独具风味
3. 豪华阁超豪华套房特设计一个适合私人聚会的就餐区
4. 柔软的沙发、以丝绸覆盖的大床、精美的刺绣，这一切都与东方主题相契合
5. 豪华阁江景房拥有落地窗和宽大而舒适的工作区域

SHANGRI-LA HOTEL, WENZHOU

温州香格里拉大酒店

Located in the centre of Central Business District of the new development area of Wenzhou, next to the International Convention and Exhibition Centre and with easy access to local government offices, Shangri-La Hotel, Wenzhou overlooks the Oujiang River and enjoys spectacular river and mountain views.

The Hotel features 409 tastefully appointed guest rooms, including 73 Horizon Club rooms and 34 suites, each with river and mountain views, and a minimum floor space of 42 square metres

— some of the most spacious rooms among Wenzhou hotels. For long-staying guests, 10 serviced apartments with fully equipped kitchenettes are also available.

The beautifully appointed Shang Palace, with 17 luxury private dining rooms, serves authentic Ou cuisine, Cantonese and Huaiyang delicacies. The trendsetting O Café restaurant offers a new style of Western buffet with Asian flavours in an open-kitchen setting.

Completion date: 2008
Location: Wenzhou, China
Designer: K.K.S Group

Photographer: Shangri-La Hotels and Resorts
Area: 60,000m²

完成时间：2008 年
项目地点：中国，温州
设计师：K.K.S 集团

摄影师：香格里拉酒店集团
面积：60,000 平方米

温州香格里拉大酒店坐落于温州滨江商务区的中心地段，毗邻会展中心，和市政府大楼相近。温州香格里拉大酒店俯瞰瓯江，坐拥青山秀水的秀丽风光。

酒店拥有409套精致的客房，其中包括73间豪华阁客房和34套套房，分别可欣赏江景和山景，房间面积在42平方米以上——是温州酒店中最宽敞的客房。为了满足长期入住客人的需要，酒店还设有10间配有全套厨房设施的酒店公寓。

优雅华丽的中餐厅香宫设有17间豪华包间供应地道的粤菜和本地淮扬菜系美食。时尚新潮的瓯咖啡可以让人领略开放式厨房概念带来的别具一格的西式自助餐与亚洲口味的完美融合。

1. The function room for government interview
2. Cocktail reception at Grand Ballroom Foyer features several chandeliers shining gleaming star lights.

1. 政府接见室
2. 大宴会厅门厅的鸡尾酒接待室以闪耀的吊灯为特色

1. Grand ballroom foyer	1. 大宴会厅前厅
2. Grand ballroom	2. 大宴会厅
3. Xin'an River room	3. 新安江
4. Feiyun River room	4. 飞云江
5. Qiantang River room	5. 钱塘江
6. Cloakroom	6. 衣帽间
7. Bridal room	7. 新娘房
8. Lift	8. 电梯
9. Yandang Mountain VIP room	9. 雁荡山贵宾厅
10. Pingyang room	10. 平阳
11. Cangnan room	11. 苍南
12. Yongjia room	12. 永嘉

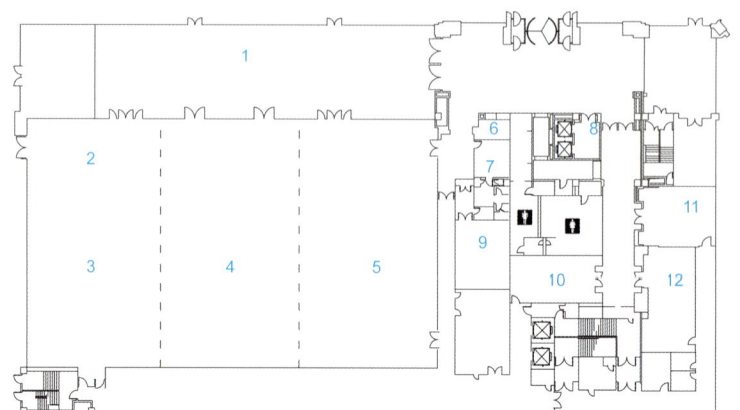

1. Private dining room in Shang Palace offers authentic Chinese dining atmosphere.
2. The interior of Shang Palace is inspired by traditional Chinese culture. Chinese painting is hanging on the wall, and screens with bamboo motif separate different areas.
3. Horizon Club Lounge is furnished with groups of sofas and comfortable chairs, offering a causal ambience.

1. 香宫中餐厅的包房营造出正宗的中式就餐氛围
2. 香宫的设计从中国传统文化中获取了灵感；墙壁上悬挂着中国画，并利用带有竹子图案的屏风来分割空间
3. 豪华阁贵宾廊内设有舒适的沙发和座椅，营造出休闲的氛围

1. Junior ballroom	1. 宴会厅
2. Nanxi room	2. 楠溪
3. Oujiang room	3. 瓯江
4. Function room foyer	4. 多功能厅前厅
5. Ballroom foyer	5. 宴会厅前厅
6. Escalator	6. 自动扶梯
7. Tai Shun	7. 泰顺
8. Business centre	8. 商务中心
9. Jinhua room	9. 金华
10. Taizhou room	10. 台州
11. Lucheng room	11. 鹿城
12. Foyer	12. 前厅
13. Wenzhou room	13. 温州
14. Lishui room	14. 丽水
15. Leqing room	15. 乐清
16. Rui'an room	16. 瑞安

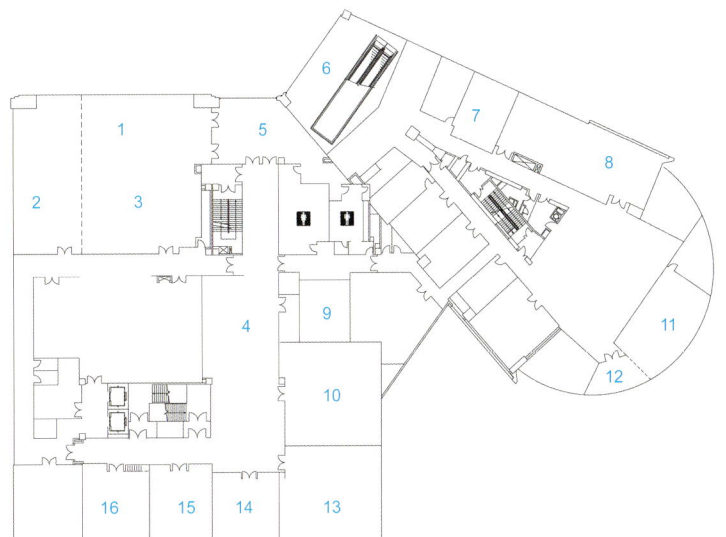

1. The trendsetting O Café restaurant offers a new style of Western buffet with Asian flavours in an open-kitchen setting.
2. With magnificent high ceiling, artistic chandeliers and painting masterpieces, Grand Ballroom represents grandeur.
3. The boardroom employs reserved colour palette creating stricter atmosphere for business meetings.

1. 时尚新潮的瓯咖啡可以让人领略开放式厨房概念带来的别具一格的西式自助餐与亚洲口味的完美融合
2. 华丽的挑高天花板、充满艺术气息的吊灯和名画让大宴会厅显得富丽堂皇
3. 会议室采用稳重的色调为商务会面添加了严肃的氛围

1. Suites feature a spacious lounge room and study area with large executive desk.
2. Deluxe Rooms offer spacious comfort with contemporary décor.
3. With an elegant deco and spacious comfort, Premier Rooms offer extra space and striking river and mountain views.

1. 套房拥有宽敞的休息室，工作区设有宽大的书桌
2. 豪华套房采用现代装饰，舒适而宽敞
3. 高级客房拥有优雅的装饰和宽敞的空间，可以俯瞰山河的美景

SHANGRI-LA HOTEL, XI'AN

西安香格里拉大酒店

Situated amidst beautifully landscaped gardens – Shangri-La Hotel Xi'an is located in the Gaoxin Hi-Tech Zone within a city rich in historical and cultural significance. The hotel is convenient to the region's many historical sites and museums – including the Terracotta Warriors Museum and Xi'an's breathtaking city walls – and is within walking distance of Century Ginwa and Golden Eagle shopping malls. Shangri-La Hotel, Xi'an offers four trendsetting designer restaurants and features the largest and most sophisticated meeting and banqueting facilities in Xi'an. And with luxurious spa treatments and a fully equipped health club with heated pool, it's the perfect place to soothe mind, body and soul.

Among the hotel's deluxe facilities are a sparkling 25-metre indoor pool, a luxurious day spa, and a fully equipped health club with Jacuzzi, steam room, and sauna. The hotel's three world-class restaurants offer a delectable selection of exquisite cuisine – while the Lobby Lounge provides an elegant backdrop for afternoon tea or evening cocktails. The property is also one of the largest and most sought-after event space in Xi'an – including three expansive ballrooms and eight additional function rooms with a full range of technological capabilities – making Shangri-La Xi'an an ideal host for conferences, weddings and other special events. Shangri-La Xi'an features 395 beautifully appointed guestrooms and suites – each with generous amenities. Spectacular floor-to-ceiling windows afford picturesque city views. Horizon Club-level accommodations provide guests with additional premium features – including a private concierge, late-checkout privileges and access to the exclusive Horizon Club Lounge with complimentary breakfast, snacks and evening cocktails.

Completion date: 2007
Location: Xi'an, China
Designer: Solari Design Limited

Photographer: Shangri-La Hotels and Resorts
Area: 20,346m²

完成时间：2007年
项目地点：中国，西安
设计师：索拉里设计公司

摄影师：香格里拉酒店集团
面积：20,346平方米

环簇在美丽的景观花园之中，西安香格里拉大酒店位于西安——这座历史丰富、文化深远的古城的高新开发区。从酒店可以轻松到达各个历史遗迹和博物馆，包括兵马俑博物馆和西安古城墙，毗邻世纪金花和金鹰商贸。西安香格里拉大酒店拥有四家创新的设计餐厅以及西安最大、最完善的会议和宴会设施。奢华的水疗服务和设施齐全的健身俱乐部（配有恒温游泳池）是舒缓身心的完美场所。

酒店拥有一个25米长的室内游泳池、一个奢华的日间水疗中心和设施齐全的健身俱乐部（配有极可意按摩浴缸、蒸汽浴室和桑拿房）。酒店的三间世界顶级餐厅提供了令人愉悦的精致美食，而大堂酒廊则适合享用下午茶或晚间鸡尾酒。酒店拥有西安最大、最受欢迎的活动空间，包括三间宽敞的宴会厅和八间设施齐全的功能厅，适合举办会议、婚礼和各色特殊活动。西安香格里拉大酒店拥有395间设计独特的客房和套房。华丽的落地窗远眺城市的如画的景色。豪华阁的住宿设施格外奢华，享有私人门房服务和延迟退房特权，还可以进入专属的豪华阁酒廊享用免费的早餐、甜点和鸡尾酒。

1. Western wedding setup in Grand Ballroom employs gold palette to show imperial luxury.
2. Chinese wedding setup in Grand Ballroom inherits Chinese red to express Chinese happiness.
3. Wedding setup in Xi'an Room

1. 大宴会厅的西式婚礼布置采用金色来展示皇室奢华
2. 大宴会厅的中式婚礼布置传承了中国红来表达喜悦
3. 西安厅的婚礼布景

1. With market style design, Yi Café is an international culinary journey, where the entire kitchen is a stage featuring show kitchens set in a lively, cosmopolitan atmosphere.

2. Tian Xiang Ge is an authentic Chinese restaurant, featuring traditional private Chinese dining rooms.

3. The private dining room in Siam Garden is specialised with Thai décor.

1. 怡咖啡厅采用市场风格，表演厨房营造出四海一家的气氛
2. 天香阁中餐厅以传统中式包房为特色
3. 暹罗花园餐厅的包房采用了泰式装饰

1. Decorated with delicate fences and lanterns, dining in Siam Garden experiences Thai lifestyle.
2. A corner in Siam Garden with gauze curtain and bamboo resembles a fairy land, echoing the meaning of the hotel's name Shangri-La.

1. 暹罗花园采用了精致的栅栏和灯笼装饰，极具泰国风情
2. 暹罗花园一角的纱帘和竹子宛如仙境，与酒店的名字"香格里拉"遥相呼应

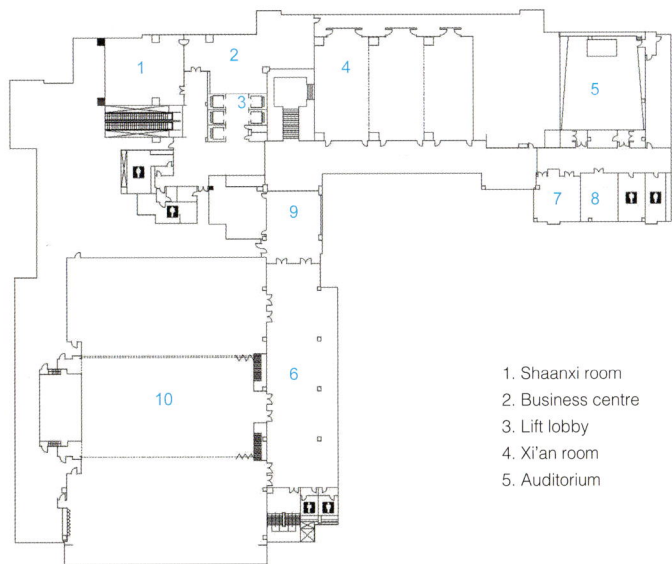

1. Shaanxi room
2. Business centre
3. Lift lobby
4. Xi'an room
5. Auditorium

6. Pre-function area
7. Function room 1
8. Function room 2
9. Function room 3
10. Grand ballroom

1. 陕西厅
2. 商务中心
3. 电梯大厅
4. 西安厅
5. 礼堂

6. 准备区
7. 功能厅1
8. 功能厅2
9. 功能厅3
10. 大宴会厅

1. The entrance to Tian Xiang Ge is inspired with traditional Chinese garden style.
2. Arch door, lacquered column, wooden lattice, red lantern and round tables compose a Chinese décor garden—Tian Xiang Ge.
3. The luxurious interior of function room is suitable for distinguished guest.

1. 天香阁的入口设计从传统中式花园中获得了灵感
2. 拱门、漆柱、木格、红灯笼和圆桌共同组成了中式花园——天香阁
3. 奢华的功能厅专为尊贵的客人设计

1. The screens in Presidential Suite add more sense of space.
2. The 228-square-metre Presidential Suite offers superlative levels of luxury and elegance with a bird's-eye view of hotel's exquisitely landscaped gardens.
3. Expansive and luxurious 114-square-metre Premier Suites are spacious and finely appointed, offering elegant views of hotel's exquisitely landscaped gardens.
4. Shangri-La Suite presents stylish luxury with its separate sitting room and dining room.
5. With magnificent views of the hotel's exquisitely landscaped gardens, Premier Suite offers luxuriously appointed accommodations, excellent amenities and modern facilities.

1. 总统套房的屏风增添了空间感
2. 228平方米的总统套房将提供极致奢华的体验，能够俯瞰酒店花园的美景
3. 114平方米的高级套房宽敞舒适而又精致典雅，享有酒店景观花园的美景
4. 香格里拉套房以独立的起居室与餐厅展现时尚奢华
5. 香格里拉套房室内面积达152平方米，宾客可从房内欣赏酒店花园壮观精美的景致。套房住宿环境豪华舒适，设施一应俱全

SHANGRI-LA HOTEL, SUZHOU
苏州香格里拉大酒店

Positioned in the Suzhou New District, the Shangri-La Hotel Suzhou stands out with its unique, futuristic design. Created to encompass extensive meeting facilities that accommodate the business clients, the hotel also supports a comprehensive list of luxury services and amenities. Furnishing beautifully designed and spacious rooms that offer the model of comfort and serenity.

Shangri-La Hotel, Suzhou offers spacious, modern living quarters with magnificent views of Suzhou. All suites include separate sitting and dining areas, large-screen televisions, DVD entertainment systems, and beautiful bathrooms stocked with designer bath amenities.

The hotel provides 390 spacious and tastefully designed rooms and suites, as well as 20 serviced apartments. This Suzhou luxury hotel is home to variety of Asian, Western, and International restaurants that each offer the highest quality ingredients and gracious Shangri-La experience. The hotel meeting facilities are preeminent in the area. The Health Club features a wide variety of indoor and outdoor recreational facilities. The 24-hour gymnasium offers cardiovascular and weight training equipment, a steam sauna and a Jacuzzi, as well as an indoor swimming pool, two tennis courts, and six spa treatment rooms. Horizon Club guests also enjoy access to the private lounge, which features double-high ceilings with spectacular views of Suzhou.

The hotel also hosts the city's best nightlife. The hotel's hotspot is Club Red, with private rooms, a full bar and an incredible dance floor all set in a fantastic interior complete with decorated walls and crystal droplights. For a quiet drink and live entertainment, the Lobby Lounge is the perfect spot.

Completion date: 2007
Location: Suzhou, China
Designer: L.R.F design

Photographer: Shangri-La Hotels and Resorts
Area: 25,000m²

完成时间：2007 年
项目地点：中国，苏州
设计师：L.R.F 设计

摄影师：香格里拉酒店集团
面积：25,000 平方米

苏州香格里拉大酒店位于苏州高新区，拥有独特的未来感设计。酒店规划建造了充足的会议宴会设施。同时支持一系列奢华服务和设施。设计美观而宽敞的客房打造了奢华而宁静的设计典范。

苏州香格里拉大酒店的套房宽敞而现代，可以俯瞰苏州全景。所有套房都拥有独立的会客和就餐区、大屏幕电视、DVD娱乐系统和美观的浴室（配有设计卫浴设施）。

酒店拥有390间宽敞舒适的客房和套房，以及20套酒店式服务公寓。酒店内设有各式亚洲、西方和国际餐厅，每家餐厅都提供高品质的美食和雅致的香格里拉体验。酒店的会议设施在该地区出类拔萃。健身中心拥有一系列的室内外娱乐设施。24小时健身房设有心血管功能和重量训练设施、蒸汽桑拿和按摩浴缸，室内泳池，两个露天网球场和6间水疗护理室。豪华阁的客人可以进出豪华阁贵宾廊，酒廊以双层挑高空间和俯瞰城市风景为特色。

酒店还拥有苏州最好的夜生活场所。红吧拥有私人包房、吧台和无与伦比的舞池。装饰墙和水景吊灯打造了酒吧美轮美奂的室内设计。而对于小酌和观看现场演出来说，大堂吧是最佳的场所。

1. The spacious and luxurious corridor
2. Horizon Club Lounge features double-high ceilings with spectacular views of Suzhou.
3. The colourful pendants hanging under the ceiling of Café Soo activate the space with single colour tone.

1. 宽敞而奢华的走廊
2. 豪华阁贵宾廊以双层挑高空间和城市风景为特色
3. 色彩斑斓的吊灯为洙咖啡厅增添了活泼的色彩

1. With live entertainment and a full range of snacks and cocktails, the Lobby Lounge Is one of the hotel's most comfortable and frequented areas.
2. Shang Palace is the ideal place to savor traditional Chinese décor.
3. The interior of Nishimura Restaurant affords sense of minimalist.

1. 大堂酒廊呈上现场表演和各色甜点、鸡尾酒，是酒店最舒适、最受欢迎的场所之一
2. 香宫中餐厅是品鉴中国传统装饰的理想场所
3. 西村日本料理餐厅采用了极简主义风格设计

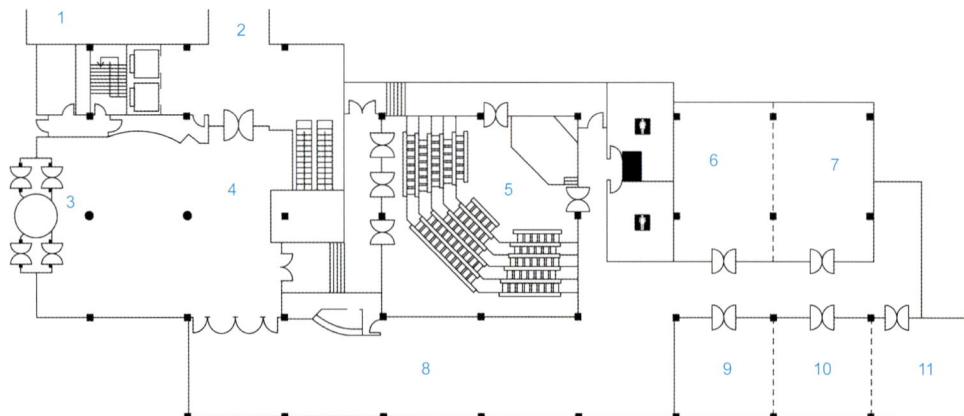

1. Hotel entrance	1. 酒店入口
2. Hotel lobby	2. 酒店大堂
3. Banquet entrance	3. 宴会入口
4. Banquet lobby	4. 宴会大堂
5. Auditorium	5. 演讲厅
6. Jian Shan	6. 见山楼
7. Yuan Xiang	7. 远香堂
8. Foyer	8. 宴会前厅
9. He Feng	9. 荷风亭
10. Zhui Yun	10. 缀云峰
11. Tong Zuo	11. 同坐轩

1

2

1. Metropolitan Ballroom offers a warm festive atmosphere, which can accommodate 220 guests for dinner banquet and 350 people for cocktail reception.
2. The five function rooms on Level One are configured in various sizes to cater to more intimate meetings or social gatherings.
3. The 1,310-square-metre pillarless Grand Ballroom, with a ceiling height of 8 metres, has a capacity of up to 800 guests for dinner banquet or over 1,200 people for cocktail reception.

1. 新地宴会厅营造出温馨的节庆气氛，可容纳220名宾客参加宴会或350名宾客参加鸡尾酒会
2. 一层的五个功能厅大小不一，适合举办各种宴会和集会
3. 1,310平方米的无柱式大宴会厅的天花板挑高8米，可容纳800名宾客参加宴会或1,200人参加鸡尾酒会

1. With separate sitting room or parlour, kitchenette and dining area, Presidential Suite delivers only the highest levels of elegance, sophistication, and luxury.

2. The lively pink flower decorates the Serviced Apartment of elegant and simple style.

3. Premier Room shows the elegance with a balance between oriental and occidental décor. The elegant Asian influences in the room design lend a sense of authenticity.

4. The living room at Executive Suite is comfortably furnished. Floor-to-ceiling windows look out to splendid views of the city.

5. Deluxe Room provides a peaceful and indulgent stay.

1. 独立客厅（或会客室）、厨房和餐厅让总统套房散发出高雅、精致、奢华的气息
2. 优雅的鲜花装饰着酒店式服务公寓，呈现出优雅而简洁的风格
3. 超豪华房展示了东西方装饰的优雅结合；高雅的亚洲风情营造出真实感
4. 行政套房的客厅配有舒适的家具；落地窗可以眺望城市的美景
5. 豪华房提供了宁静而舒适的居住环境

SHANGRI-LA HOTEL, GUILIN
桂林香格里拉大酒店

Adopting Chinese classic design, Shangri-La Hotel Guilin is located by Li River and covers 48,000 square metres, making it an ideal location for business, leisure and gathering activities. It accommodates a full range of spacious guestrooms with a panoramic view of fascinating Li River and landscaping there. Behind the façade of the seven-storey building with a pavilion-style roof is a thoroughly modern hotel with landscaped grounds, gourmet cuisine, and the city's largest ballroom. Design elements of the 449 rooms at the Shangri-La Hotel, Guilin include gold-and-black Chinese-style wall murals over the bed, calligraphy artwork, and rich mahogany furnishings. The beds are king-sized and topped with down duvets.

The hotel also incorporates Shangri-La's Horizon Club guestrooms and lounge, a heaven where today's traveller can experience an even higher standard of accommodation and personalised service complemented with a host of exclusive privileges and amenities. Shangri-La Hotel, Guilin has several restaurants including the casual Li Café where an international buffet-style is served at breakfast and dinner. The dramatic Lobby Lounge with nearly 30-foot-tall windows is where couples can have afternoon tea or a martini. For meetings, incentives, conventions and exhibitions, the 1,800-square-metre Grand Ballroom, the largest ballroom in Guilin, is the perfect venue for hosting high-profile events of up to 2,000 people.

Completion date: 2010
Location: Guilin, China
Designer: Wong & Tung

Photographer: Shangri-La Hotels and Resorts
Area: 48,000m²

完成时间：2010 年
项目地点：中国，桂林
设计师：王董集团

摄影师：香格里拉酒店集团
面积：48,000 平方米

1. The Chinese character for "U" means "relaxed, free and easy" which is exactly what the establishment inspires in guests of U Bar with its comfortable décor.
2. In the modern, relaxed and comfortable environment of Li Café, five live cooking stations, including tandoori and teppanyaki stalls, showcase the best of international cooking styles.
3. Classroom setup in Grand Ballroom

1. "悠"意味着"放松、自由和简单"，这正是悠酒吧的设计灵感
2. 漓咖啡餐厅的环境现代、放松而舒适，拥有五个现场烹饪的档口，其中包括印度唐杜里泥炉烧烤和日式铁板烧，展现了多样化的国际烹饪风格
3. 大宴会厅的课桌式布置

桂林香格里拉大酒店采用了中式经典设计，毗邻漓江，总面积48,000平方米，是商务、休闲和集会活动的理想场所。酒店拥有一系列宽敞豪华的客房及套房，享有漓江及其周围景观的迷人景色。 在古典宫殿造型的屋檐下，这座楼高七层的建筑内部是一家现代酒店。酒店四周环绕着山水景观，供应美食美酒，并且拥有全市最大的宴会厅。桂林香格里拉大酒店的449间客房设计典雅独特，金色和黑色相结合的中式床头水墨壁画及丰富的红木家具。宽敞舒适的床上铺有柔软的羽绒被。
酒店还设有豪华阁客房及酒廊，是旅客体验更高层次住宿和个性化服务的天堂。桂林香格里拉大酒店的漓咖啡在时尚轻松的环境中为客人提供各式国际化佳肴。大堂酒廊拥有9米高的落地窗，适合情侣及朋友品尝下午茶和鸡尾酒。1,800平方米的大宴会厅是桂林最大的宴会厅，适合举办各种会议、宴会、颁奖和展览活动，最多可容纳2,000人。

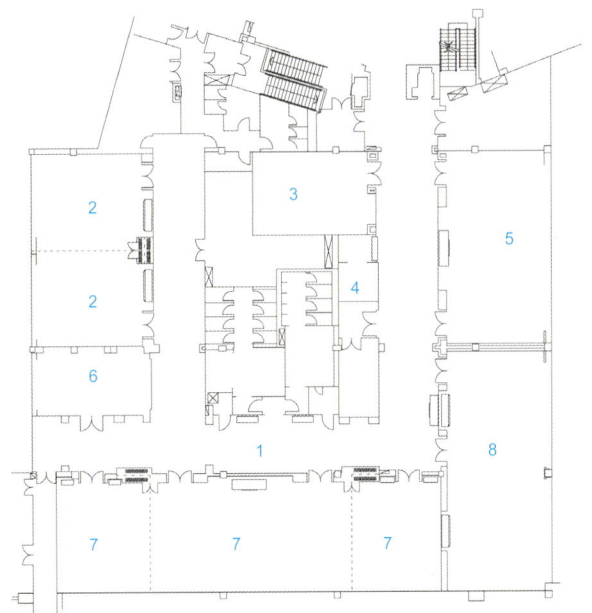

1. Foyer	1. 前厅
2. Guilin room	2. 桂林厅
3. Lipu room	3. 荔浦厅
4. Cloakroom	4. 衣帽间
5. Yangshuo room	5. 阳朔厅
6. Longsheng room	6. 龙胜厅
7. Nanning room	7. 南宁厅
8. Beihai room	8. 北海厅

1. For meetings, incentives, conventions and exhibitions, the 1,800-square-metre Grand Ballroom, the largest ballroom in Guilin, is the perfect venue for hosting high-profile events of up to 2,000 people. The image shows Chinese wedding set-up in Grand Ballroom.

2. Home to the city's largest five-star conference and meeting space, Shangri-La Hotel, Guilin, offers 4,389 sqm of elegant function rooms. All conference and function rooms feature state-of-the-art audiovisual equipment. The image is Western wedding set-up in meeting space.

1. 1,800平方米的大宴会厅能够举办会议、颁奖、会议及展示活动，可容纳2,000人；本图为中式婚礼布置

2. 桂林香格里拉大酒店拥有桂林最大的五星级会议及宴会设施，总面积达4,389平方米；所有会议室和多功能厅都拥有最先进的视听设备；本图为西式婚礼布置

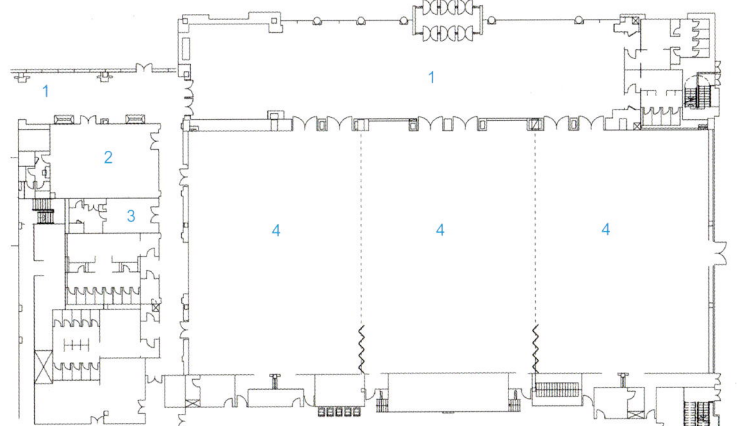

1. Foyer
2. VIP room
3. Bridal room
4. Grand ballroom

1. 前厅
2. 贵宾厅
3. 新娘化妆间
4. 大宴会厅

1. The attention is focused on the mahogany furnishings, presenting Chinese traditional luxury.

2. Design elements of the 449 rooms at the Shangri-La Hotel, Guilin include gold-and-black Chinese-style wall murals over the bed, and calligraphy artwork.

3. With marvelous views of the Li River, the Presidential Suite offers impeccable levels of luxury, with fine art pieces accentuating elegant design.

1. 红木家具是其亮点，展现了中式传统奢华

2. 桂林香格里拉大酒店的449间客房设计中融入了金色和黑色相结合的中式床头水墨壁画工艺

3. 总统套房尽揽风光旖旎的漓江美景，为宾客呈献尽善尽美的奢华享受。房间内摆设有精美的艺术品，充分彰显高雅品位

SHANGRI-LA HOTEL TOKYO
东京香格里拉大酒店

The Shangri-La Tokyo design consists of the rich and luxurious DNA of Shangri-La Hotel brand emotion to which contemporary details have been added to create the unique Shangri-La Hotel Tokyo hotel experience.

Porte Cochere – 31m long x 5.5m wide coffer with hanging crystals and beautiful marble walls welcome guests with luxurious atmosphere and stunning water feature. Symbolic back-lit carved metal art screen appears at columns in addition to warm tone marble columns detailing the back water wall across from the entry façade. Also the dark beautiful grain marble contrasts with the signature hotel graphics which introduce a dramatic entrance experience and emphasises the Shangri-La Tokyo identity.

Lift Lobby (from the basement and ground floor) – Elegant patterned custom metal lift doors by an international artist is an eye-catching element, making guests excited to enter and experience the lift ride to the public area of the hotel, on the 26th-28th floor. The lobby is on the 27th floor, restaurants and spa are on the 28th floor, and guestrooms are on 31st-36th floor. A focal table with gorgeous floral arrangement (ground floor) complements the hand-tufted area rug adds residential feeling.

Pedestrian Entry – Arriving from cold monotone architectural finishes of the outside building, the guest's eye is caught by the

5m tall large piece of art work above the glass door façade inside the vestibule. This identifies the hotel's presence.

Shuttle Lift Car – 3.5m high ceiling with crystal chandelier surrounded by highly figured rich Movingi wood panelled walls and bronze mirror. This rich experience is accented with clear acrylic handrails filled with Swarovski crystals. Rich hand-tufted carpet completes this luxurious experience.

Shuttle lift lobby, Front Desk area – Completing the front desk is a gold-leaf coffer ceiling with gorgeous art work pattern, and the front desk is made of a portoro stone slab finished off with beautiful bouquets of flowers at either side.

Dividing the front desk and lobby bar area a focal decorative glass screen screens the lounge area, which adds the contemporary elements to the space while lending a sense of intimacy and privacy to the area.

Lobby Lounge – The lounge consists of an eclectic mix of seating groups, functioning for both high tea service and evening cocktails. The area has windows to Tokyo on three sides, giving the guest a 180 degree view of the Tokyo skyline. High windows accentuate the view which is framed with dramatic drapery. The rich palette in the handmade area rug along with the silk, velvet and other rich fabrics give the guest a rich tactile experience not

Completion date: 2009
Location: Tokyo, Japan

Designer: HBA/Hirsch Bender Associates
Photographer: HBA

完成时间：2009 年
项目地点：日本，东京

设计师：HBA
摄影师：HBA

1. Completing the front desk on level 28 is a gold-leaf coffer ceiling with gorgeous art work pattern; the front desk is made of a portoro stone slab finished off with beautiful bouquets of flowers at either side.
2. Meaning "pleasure" in Italian, Piacere succeeds in the style, passion and creativity of its interior design.

1. 28层的前台区域采用了金箔天花板和艺术品装饰；前台由黑金花石板制成，两侧摆放着美丽的花束
2. Piacere在意大利语中意为"愉悦"，贯穿了室内设计的风格、激情和创造力

found in any other hotel. There is a mix of seating, from dining to cocktail to lounge to bar seating which creates a wide selection for the guest's use. Rich materials at the bar of onyx and gold leaf along with a signature custom-designed chandelier consisting of Czechoslovakian crystal finish off this amazing area.

Typical Guestroom – The guestrooms and smaller suites of the hotel have been custom-designed to create a one-of-a-kind feeling. The entry is accented with a custom-designed welcome console which creates a very residential feeling. A custom-designed closed in rich figured Anigre veneer continues the rich level of detail. The bath is unique in both size and style for Tokyo. A larger than usual bathing area of an oversize shower and bath tub is accented with marble floors and walls along with custom bath fixtures. A custom wood vanity, backlit vanity mirror with built-in TV and compartmentalized toilet complete this area. The bed chamber has rich figured wood veneer, custom built-in sofa at the window in rich velvet, a large working desk area and a large 42" diagonal wall mount TV with DVD and CD accessories. A custom mini-bar and Tea Service area are at the entry to the bedchamber along with a large oversize built-in luggage bench.

1. Common lobby 1. 大堂
2. Hotel Signage by others 2. 酒店引导标识
3. Vestibule 3. 前厅
4. Shuttle lift lobby 4. 电梯门厅
5. Entry 5. 入口

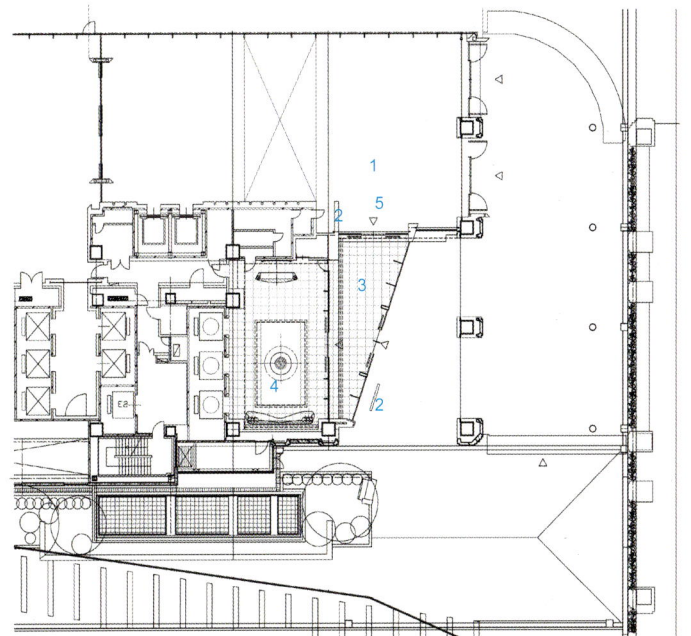

1. Shangri-La has created an exclusive retreat – Horizon Club Lounge.
2. The magnificent Shangri-La Ballroom with its gorgeous central chandelier can accommodate up to 270 guests for theatre style, up to 200 guests for dinner style, up to 250 guests for cocktail reception style arrangements.
3. Three function rooms are also available for smaller gatherings. The Conway is one of them featuring refined interior design.

1. 香格里拉的专属空间——豪华阁贵宾廊
2. 华丽的香格里拉宴会厅配有绚丽的中央吊灯，作为剧场可容纳270人，作为宴会厅可容纳200人，而作为鸡尾酒接待厅可容纳250人
3. 三个功能厅适合举办小型聚会；其中，康韦厅以精致的室内设计为特色

东京香格里拉大酒店的设计富有香格里拉酒店品牌特有的奢华元素，同时也添加了现代设计，形成了东京香格里拉大酒店的独特体验。

车辆门廊——31米长，5.5米宽的门廊内采用了悬挂水晶吊灯和美丽的大理石墙面，与绝妙的水景设施一起迎接着宾客进入奢华的氛围。柱子上象征式的背光金属雕刻艺术屏风为大理石柱增添了温暖的色调，与入口对面的水墙相得益彰。黑纹大理石与酒店的特色图案形成了对比，为东京香格里拉大酒店打造了戏剧性的入口体验。

电梯大厅（从地下室到一层）——带有典雅图案的定制电梯门由国际艺术家设计，拥有吸引眼球的元素，让客人高兴地通过电梯到达酒店27层到29层之间的公共区域。酒店大堂设在28层，餐厅和水疗中心设在29层，而客房则设在31层到36层。一楼前台上摆放着华丽的插花作品，而手工地毯区域则增添了居家感。

行人入口——从冰冷的建筑外部进入酒店，客人的眼球会立即被5米高巨型艺术品所吸引。这标志着酒店的存在感。

电梯厢——电梯厢3.5米高的天花板和水晶吊灯，四周环绕着木镶板墙和铜镜。简洁的亚克力栏杆上镶满了施华洛奇水晶。质感丰富的手织地毯进一步完善了这种奢华体验。

电梯和前台接待区——金箔天花板上带有辉煌的艺术图案，而前台则由波多若斯石板制成，两侧装饰着美丽的花束。

分割前台和大堂酒廊区的中央玻璃屏风将酒廊遮挡了起来，为空间增添了现代元素和私密感。

大堂酒廊——酒廊由各种各样的座位区组成，提供下午茶和晚间鸡尾酒服务。这一区域三面都是窗户，提供了180度的东京天际线美景。高窗的两侧挂有戏剧化的帷帐。手工装饰地毯与丝绸、天鹅绒和其他织物上丰富的图案为客人带来了丰富的触感，这在其他酒店是找不到的。酒廊设有各种座位，适合就餐、鸡尾酒、休息等各种活动，为客人提供了多种选择。玛瑙和金箔制成的吧台与定制捷克水晶吊灯共同装饰了这个迷人的空间。

标准客房——酒店客房和小型套房拥有独一无二的定制设计。入口处特别设计的迎宾台极富家居感，采用了质感丰富的安利格薄板。浴室在规模和风格上都独具特色。更大的洗浴区配有特大号花洒和浴缸，配合大理石地面、大理石墙面和定制的卫浴设施。定制的木梳妆台、背光镜、嵌入式电视和独立抽水马桶完善了这一区域。卧室装饰着丰富图案的木板、定制嵌入式天鹅绒窗边沙发、大型书桌和巨大的42寸电视（配有DVD和CD设备）。卧室门口定制的迷你吧和茶水间旁设有一个巨型的嵌入式行李长椅。

2

3

1

1. The heated indoor swimming pool completes with stunning ceiling with the motif of weave.

2. The entry is accented with a custom-designed welcome console which creates a very residential feeling. The living room of Presidential Suite features opulent luxurious details. With elegant appointments and spatial freedom, the suite was designed with total comfort in mind.

3. The bed chamber has rich figured wood veneer, custom built-in sofa at the window in rich velvet. The guestrooms and smaller suites of the hotel have been custom designed to create a one of a kind feeling.

1. 室内恒温游泳池的天花板上装饰着波纹图案

2. 入口独特的接待台营造出居家感；总统套房的客厅以丰富奢华的细节设计为特色；套房拥有典雅的设施和宽敞的空间，十分舒适

3. 卧室窗台上设有定制的嵌入式沙发；客房和小型套房的定制设计打造了独一无二的感觉

SHANGRI-LA'S FAR EASTERN PLAZA HOTEL, TAINAN

香格里拉台南远东国际大饭店

An Amazing Welcome

As a five-star hotel located in Tainan, a small city with rich cultures, the design of it needs to achieve a balance between gorgeous refinement and warm welcome. Although the hotel's scale and materials carry on Shangri-La's traditional glamour, the design team apply various techniques to make the interior space more human. With a warm feeling, the local wood eases the hotel's hollowness and echoes the local lavish landscapes. And the simple and interesting texture of wood will add to the comfortable atmosphere. The designers intentionally avoid excessive luxury; therefore the designers choose a series of normal materials and give full play to their textures and uses.

Since Tainan is located on the seismic belt, firm beams and columns are the necessary. AB Concept wraps them with oak and gives each one a different texture. For instance, the column in the centre of the lobby is transformed into a trendy floor lamp with stream lines. Though in a large scale, the wood, the upholstery and the granite water wall with air illumination add warmth to the lobby.

The designers use various interesting methods to break the large space into different parts: they put the reception desk in the niche to create a more approachable space scale and sense of intimacy. In addition, the back-lit screen will add layers to the space while the handrails with beautiful lines will line out individual spaces for seating. Although the hotel's proportions and luxury show a quite lush life, the comfort is the priority. Therefore, the hotel uses chairs and floor lamps with Asia characters, and encourage the guests to stay longer with a home-style design.

The café diffuses a sense of leisure and creation. The designers again use relaxing dining atmosphere to balance the hotel's magnificence. The trees and canvas frames create a sense of security for the large atrium, while the rich sunset glow colours and lively decorative lightings activate the whole atmosphere. Taking "Food Stage" as a concept, the designers attract the guests' eyes to different cook stations. The custom-designed screen takes Japanese bento as its design concept, allowing a certain interaction between different areas. But the design team realise the importance of zoning for such a large restaurant and

Completion date: February, 2009
Location: Tainan, China
Designer: Ed Ng, Terence Ngan (AB Concept)

Photographer: Chester Ong, AB Concept
Area: 118,082m²

完成时间：2009 年 2 月
项目地点：中国，台南
面积：118,082 平方米

设计师：伍仲匡，颜学添（奥必概念）
摄影师：切斯特·昂格；奥必概念

create a special way for guests to return to their seats.

The Chinese restaurant is furnished with polished wood and silk fabrics, modest and luxurious. With the advantage of being South Taiwan's highest architecture, the restaurant enjoys an unobstructed view. The designers take the VIP suite as its inspiration and wrap the large round tables with heavy bronze bead curtains. In this way, the guests can both enjoy the beautiful view and privacy. At daytime, the restaurant with white jade tones looks bright and clean; at night, the soft atmosphere created by light will immediately turn this place into a unique place for delicious cuisines.

令人惊叹的宾至如归

这座五星级酒店位于台南这个文化味道浓厚的小城市。在设计方面需要在华丽精致与宾至如归之间取得平衡。尽管酒店的规模与建筑物料一如既往地尽显香格里拉品牌的魅力，设计团队采取了多项技巧，令室内空间更加人性化。本地木材洋溢的温暖感，缓和了酒店硕大的空旷感，并且暗合当地的青翠景色。而木材简朴而有趣的质感则添上舒适的气氛。设计师特意避免过分奢华，因此选用了一系列寻常物料，并充分发挥其质感及功用。

由于台南位处地震带，牢固的梁柱是必须之物。AB Concept将它们全部以橡木包装起来，然后赋予每一根梁柱不同的肌理，例如位于大堂中央的一根，便化身成一尊流线型的时尚座灯。尽管面积偌大，大堂的木材、衬垫及由空中照明的花岗石流水墙令空间洋溢暖意。

设计师采用不同有趣方法将硕大的空间化整为零：他们将接待处置于壁龛中，营造更平易近人的空间比例与亲切感。此外，又以背后投射光源的屏风为空间增加层次感，并以充满线条美的栏杆划出独立的位置摆放座位。尽管酒店的面积和奢华很容易令人意乱情迷，不过舒适才是首要考虑。因此，酒店用上充满亚洲风格的椅子及座灯，并特意以家居风格设计，以鼓励宾客流连徜徉。

咖啡室里散发着休闲创意，设计师再一次以悠闲用膳的气氛平衡酒店建筑的慑人气势。偌大的中庭以大树和帆布架营造安全感，再配以丰富的晚霞色彩及活泼的装饰照明，令整个空间跃动起来。设计师以"食物舞台"为概念，吸引宾客的目光穿梭于各个烹调站之间。以日式便当为设计理念的特制屏风，容许各个区域之间一定程度的互动。不过设计团队亦深谙区域划分对庞大如斯的餐厅极其重要，故以特别方法协助宾客轻易返回自己座位。

中餐厅以经打磨的木材和生丝织品装潢，散发低调奢华，并充分发挥身处台湾南部最高建筑物顶楼的优势，坐拥全无遮挡的景致。设计师以私家贵宾房的概念，将多张大型圆桌以厚实的铜珠帘包围起来，让宾客既得享景色，又享有一定程度的私隐。在日间，白玉色调的餐厅光洁明亮，到了入黑之后，灯光营造的柔和气氛，令此处迅即成为区内最具特色的美膳胜地。

1. The view from Café at Far Eastern towards top of the hotel
2. The detail of the lobby
3. The trees and canvas frames create a sense of security for the large atrium.

1. 从咖啡室看酒店顶部
2. 大堂细部设计
3. 偌大的中庭以大树和帆布架营造安全感

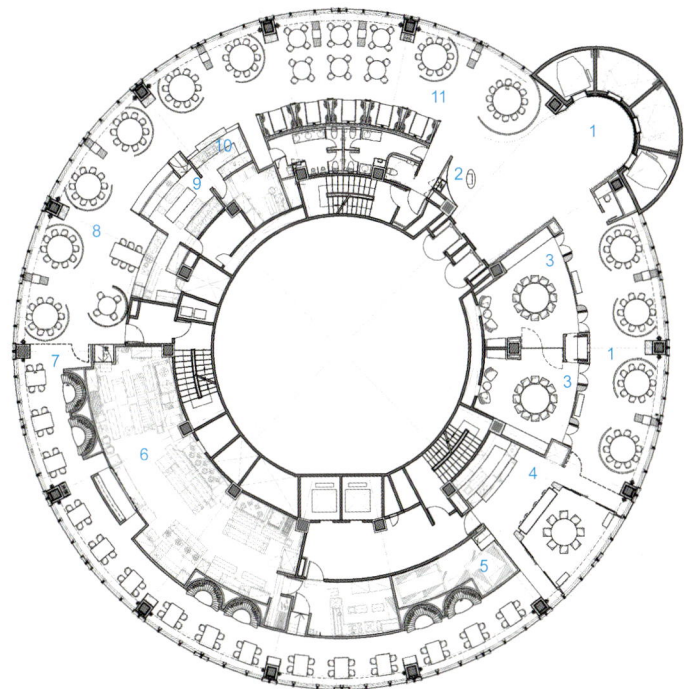

1. Foyer	1. 门厅
2. Reception counter	2. 接待柜台
3. VIP room	3. 贵宾房
4. Tea Bar	4. 茶水吧
5. Freezer room	5. 冰柜室
6. Kitchen	6. 厨房
7. Dining along the windows	7. 窗边用餐区
8. Semi-open room	8. 半开放式房间
9. Show kitchen	9. 展示式厨房
10. Water Bar	10. 水吧
11. Decks	11. 卡座

1. The reception desk is put in the niche to create a more approachable space scale and sense of intimacy.
2. In Café at Far Eastern, the rich sunset glow colours and lively decorative lightings activate the whole atmosphere.
3. Shanghai Pavilion is furnished with polished wood and silk fabrics. The designers take the VIP suite as its inspiration and wrap the large round tables with heavy bronze bead curtains.
4. The design team realise the importance of zoning for such a large restaurant and create a special way for guests to return to their seats in Café at Far Eastern.

1. 接待处被置于壁龛中，营造更平易近人的空间比例与亲切感
2. 咖啡室里，丰富的晚霞色彩及活泼的装饰照明，令整个空间跃动起来
3. 中餐厅以经打磨的木材和生丝织品装潢；设计师以私家贵宾房的概念，将多张大型圆桌以厚实的铜珠帘包围起来
4. 设计团队深谙区域划分对庞大如斯的餐厅极其重要，故以特别方法协助宾客轻易返回自己座位

1. Shanghai Pavilion is furnished with polished wood and silk fabrics, modest and luxurious.
2. The custom-designed screen takes Japanese bento as its design concept, allowing a certain interaction between different areas.
3. At daytime, Shang Pavilion with white jade tones looks bright and clean.

1. 中餐厅以经打磨的木材和生丝织品装潢，低调奢华
2. 以日式便当为设计理念的特制屏风，容许各个区域之间一定程度的互动
3. 在日间，白玉色调的餐厅光洁明亮

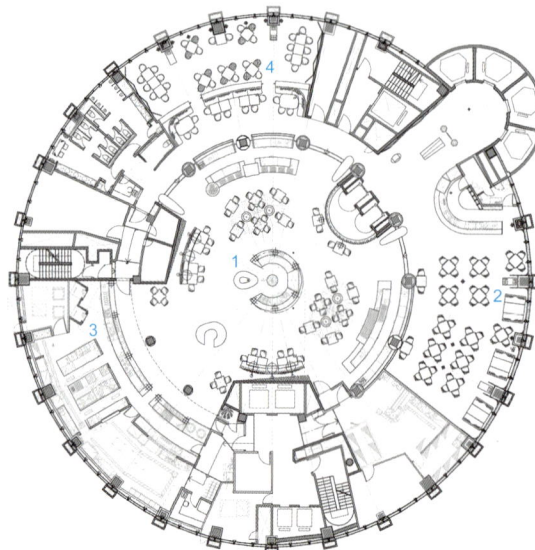

1. Atrium dining area
2. Dining area along the windows
3. Show dining area
4. Terrace dining area along the windows

1. 中庭用餐区
2. 窗边用餐区
3. 展示式用餐区
4. 窗边高平台用餐区

1. The hotel offers a distinctive range of exciting wedding venues. The image presents ballroom with wedding setting.

2. The hotel features Tainan's only amphitheatre-style auditorium. This sophisticated facility is complemented by an elegant ballroom capable of holding up to 1,300 people.

1. 酒店提供各色别具匠心的婚礼场地。图片展示了宴会厅礼堂的婚宴布置

2. 酒店拥有台南独一无二的阶梯式礼堂。该多功能设施拥有一个可容纳1,300人的高雅宴会厅

KERRY HOTEL PUDONG, SHANGHAI
上海浦东嘉里大酒店

Located in the new Kerry Parkside complex in the heart of Pudong, the hotel marks the debut of a new five-star hotel brand by Shangri-La Hotels and Resorts: Kerry Hotels.

Soaring 31 storeys over the Kerry Parkside complex, Kerry Hotel Pudong, Shanghai offers 574 guestrooms and suites, ranging in size from 42 to 168 square metres, with sweeping views of Century Park or the city. Sleek and contemporary furnishings provide optimum comfort for work, relaxation or socialising. Sliding wall panels partition the guestrooms, leading to separate bath and shower facilities. Seven floors are devoted to Club accommodation, including 33 suites. All suites have a separate living room providing a residential-style ambience.

Kerry Hotel Pudong, Shanghai introduces one of Shanghai's most innovative dining and entertainment concepts, an integrated three-in-one designer showcase developed by boutique design consultancy Stickman Tribe – The COOK, The MEET and The BREW.

The COOK, the main restaurant with 11 live show kitchens and a gourmet delicatessen has seating for 365 people, including an outdoor patio. The COOK's interior showcases a bold mix of traditional and modern materials and finishes, such as handwritten menus on rustic wooden blackboards, Corian marble countertops mixed with rustic crate-stamped timber drop ceilings and custom-designed tiles. All design elements blend together to produce an eclectic lifestyle theme that reflects a bustling deli-style marketplace in contemporary Shanghai.

Evoking industrial chic, The BREW's interiors revolve around the "flying brewery", custom designed and built three-storey stainless steel vats used for brewing and storing its signature brews. Forming a dramatic visual centrepiece and framing the brewery's steel vats in the background is an eye-catching, multi-tiered steel and glass chandelier over the centre of the bar that conveniently doubles as a beer pint glass rack.

Diners at The MEET, a contemporary steakhouse and speciality grill opening in April 2011, are greeted by a theatrically lit "Ageing Room" window dominating the restaurant entrance. A scarlet-hued

Completion date: February, 2011
Location: Shanghai, China
Designer: Kohn Pederson Fox Associates PC and
Aedas Ltd.; ARA Design; Stickman Tribe Design

Photographer: George Mitchell, Michael
Weber, Scott Wright, Todd Anthony Tyler
Area: 70,000m²

完成时间：2011 年 2 月
项目地点：中国，上海
设计师：KPF 建筑事务所；凯达环球有限公司；
ARA 设计；斯迪克曼·特来布设计

摄影师：乔治·米切尔；麦克·韦
伯；斯科特·怀特；陶德·安
东尼·泰勒
面积：70,000 平方米

vertebrae-inspired light fixture stretches down the expanse of the restaurant and makes a dramatic visual centrepiece to the restaurant and counterpoint to the gleaming tops of The BREW's glass encased brewery vats below.

The hotel houses the city's largest hotel-based sports club with Kerry Sports, a recreation destination spanning 6,000 square metres over three floors. The facilities include a 24-hour gymnasium with studios for aerobics, spinning, Pilates and yoga (including hot yoga), a 25-metre heated indoor swimming pool, an outdoor basketball court and tennis court and an outdoor roof garden with a jogging track. The 700-square-metre Adventure Zone, an indoor children's fantasy playground, features a two-storey fun slide, age-appropriate play "zones" and three festive themed party rooms, ideal for birthday celebrations.

Kerry Hotel Pudong, Shanghai unveils the largest portfolio of hotel meeting and banqueting facilities in the city, with more than 7,300 square metres of space. Dominating the hotel's third floor are two ballrooms, the 2,230-square-metre Grand Shanghai Ballroom and the 1,018-square-metre Pudong Ballroom, which can be interconnected. The Grand Shanghai Ballroom is pillar-free and can accommodate more than 2,800 people theatre style and 1,600 guests for banquets. With a ceiling soaring up to nine metres and direct lift access from the car park, the ballroom can easily be transformed into a car showroom for model launches. In addition, the hotel features 26 multi-purpose function rooms, most with natural daylight.

1. Forming a dramatic visual centrepiece and framing the brewery's steel vats in the background is an eye-catching, multi-tiered steel and glass chandelier over the centre of the bar that conveniently doubles as a beer pint glass rack.

2. The private dining room at Blossoms employs bold colour palette and luxurious furnishing, creating prosperous tone.

3. With appropriate layout of the furniture, Plum and Orchid private dining room at Blossoms combines oriental and occident furniture harmoniously.

1. 形成戏剧化的视觉中心并围绕起啤酒厂钢桶的一个引人注目的多层钢铁玻璃吊灯，吊灯同时还可以用作一个啤酒杯架

2. 百花万包房采用了大胆的色彩搭配和奢华的装饰，显得雍容华贵

3. 梅兰包房的家具搭配完美地结合了东西方特色

上海浦东嘉里大酒店位于浦东嘉里城，是香格里拉酒店集团旗下首个全新的五星级嘉里酒店品牌。

31层高的浦东嘉里大酒店拥有574间客房和套房，面积从42到168平方米不一，俯瞰着世纪公园或上海城市的美景。时尚而现代的装饰为工作、休闲或社交提供了极致舒适的空间。客房的滑动壁板隔断通往独立浴缸和淋浴设施。酒店有七层嘉里阁行政楼层客房，包含33间套房。所有套房都拥有独立的客厅，营造出家居感。

上海浦东嘉里大酒店引入了上海最具创新的餐饮娱乐概念，采用了由设计咨询公司斯迪克曼·特来布所设计的三位一体餐饮设施——The COOK·厨餐厅、The MEET·聚扒房和The BREW·酿啤酒坊。

The COOK·厨餐厅是一家拥有11间开放式厨房的全日制餐厅，可同时为365人提供美食服务，还设有一个露天平台。餐厅的室内设计大胆地融合了传统和现代的材料和装饰，例如粗糙木黑板上的手写菜牌、可丽耐大理石台面与纯朴的木板条吊顶和定制瓷砖相互结合。所有设计元素都被融合在一起，营造出中性的主题，反映了当代上海繁华的餐饮市场。

The BREW·酿啤酒坊的室内设计充满了工业感，围绕着"飞行的啤酒厂"这一主题展开，餐厅特别设计一个三层不锈钢大桶来酿造并存储特色啤酒。一个引人注目形如啤酒桶的多层钢玻璃吊灯成为了餐厅的视觉焦点，吊灯同时还可以用作一个啤酒杯架。

The MEET·聚扒房是一家兼具现代与时尚休闲的牛排馆和烧烤餐厅，开放于2011年4月。餐厅入口的可视烤肉房令人惊叹。猩红色的脊椎形灯具纵贯整个空间，形成了餐厅的视觉中心，与The BREW·酿啤酒坊的玻璃啤酒桶闪耀的顶部形成了呼应。

酒店拥有上海最大的酒店运动俱乐部——嘉里健身。三层的健身中心总面积为6,000平方米，包括24小时健身房（设有有氧运动和动感单车教室）、露天篮球场、网球场和带有慢跑道的户外屋顶花园。700平方米的冒险区专为儿童设

计，以两层楼高的滑梯、适龄游乐区和三个主题派对房（十分适合生日庆典）组成。

上海浦东嘉里大酒店拥有上海最大的酒店会议和宴会设施，总面积超过7,300平方米。2,230平方米的大上海宴会厅和1,108平方米的浦东宴会厅相互连通，占据着酒店的三楼。大上海宴会厅采用无柱式设计，可容纳2,800人与会或1,600人就餐。高达9米的宴会厅配有直达停车场的电梯，使其可以轻易变成一个汽车展室。此外，酒店还拥有26间多功能厅，大多数都享受了自然光照。

1. A scarlet-hued vertebrae-inspired light fixture stretches down the expanse of the restaurant and makes a dramatic visual centrepiece to the restaurant and counterpoint to the gleaming tops of The BREW's glass encased brewery vats below.
2. The hotel's main a la carte restaurant with 11 live designer theatre kitchens and a gourmet delicatessen, The COOK offers a feast for the senses.
3. Corian marble countertops mixed with rustic crate-stamped timber drop ceilings and custom-designed tiles. In the COOK, all design elements blend together to produce an eclectic lifestyle theme that reflects a bustling deli-style marketplace in contemporary Shanghai.

1. 猩红色的脊椎形灯具纵贯整个空间，形成了餐厅的视觉中心，与The BREW·酿啤酒坊的玻璃啤酒桶闪耀的顶部形成了呼应
2. The COOK·厨餐厅是一家拥有11间开放式厨房的全日制餐厅，可同时为365人提供美食服务
3. The COOK·厨餐厅里，可丽耐大理石台面与纯朴的木板条吊顶和定制瓷砖相互结合；所有设计元素都被融合在一起，营造出中性的主题，反映了当代上海繁华的餐饮市场

1. Simply relax at the outdoor terrace of The BREW, taking in views of Century Park.
2. The private dining room in The MEET

1. 在The BREW·酿啤酒坊的露天平台放松休息，遥看世纪公园的风景
2. The MEET·聚私人包厢

1. Private dining room
2. PDR restroom
3. Corridor
4. To 2nd floor lobby
5. Male restroom
6. Female restroom

1. 私人就餐间
2. PDR休息室
3. 走廊
4. 往二层大厅
5. 男洗手间
6. 女洗手间

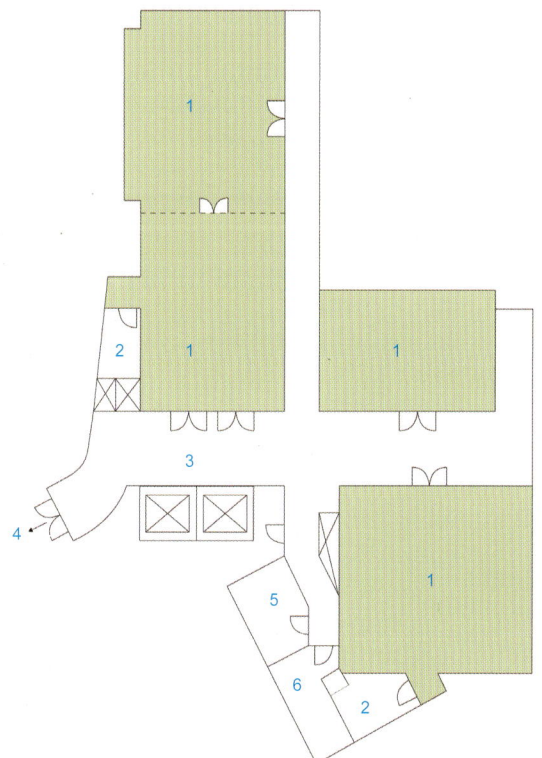

1. Dominating the hotel's third floor are two ballrooms, the 2,230-square-metre Grand Shanghai Ballroom and the 1,018-square-metre Pudong Ballroom, which can be interconnected.

2. The Grand Shanghai Ballroom is pillar-free and can accommodate more than 2,200 people theatre style and 1,600 guests for banquets. With a ceiling soaring up to nine metres and direct lift access from the car park, the ballroom can easily be transformed into a car showroom for model launches.

1. 2,230平方米的大上海宴会厅和1,108平方米的浦东宴会厅相互连通，占据着酒店的三楼

2. 大上海宴会厅采用无柱式设计，可容纳2,200人与会或1,600人就餐；高达9米的天花板和停车场直达电梯让宴会厅可以轻易变成一个汽车展室

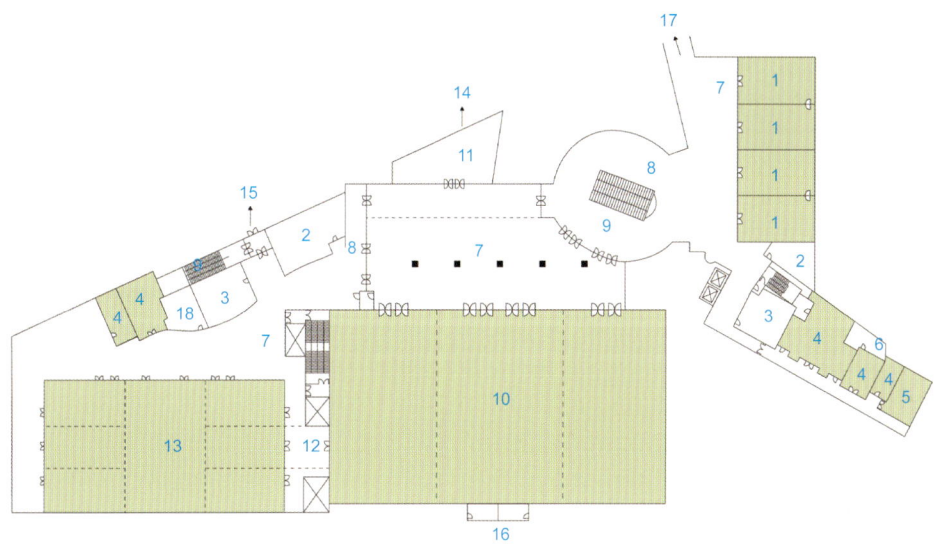

1. Function room	10. Grand Shanghai ballroom	1. 多功能厅
2. Female restroom	11. Garden balcony	2. 女洗手间
3. Male restroom	12. Connection to grand ballroom	3. 男洗手间
4. VIP room	13. Pudong ballroom	4. 贵宾厅
5. Board room	14. View to courtyard	5. 董事会议厅
6. Restroom	15. To office tower meeting centre	6. 休息室
7. Pre-function area	16. Changing room	7. 迎宾区
8. Foyer	17. To the club lounge; to the business centre and office centre	8. 前厅
9. Escalator	18. Cloaks room	9. 电梯

10. 上海大宴会厅
11. 露天花园
12. 连接大会议厅
13. 浦东大宴会厅
14. 面向庭院
15. 往办公大楼会议中心
16. 更衣室
17. 往嘉里阁贵宾廊，办公及商务中心
18. 衣帽间

1. Swimming Pool Jacuzzi is decorated with blue pillows, echoing the blue water in the swimming pool and creating a chic ambience.
2. The prosperous colour palette creates luxurious atmosphere for gathering.
3. Guests may relax with a drink on pony hair and leather armchairs near a fireplace flanked by hand-carved wooden bullhorns in The MEET.

1. 极可意按摩浴池四周点缀着蓝色的抱枕，与池中的水色相呼应，营造出时尚的氛围
2. 富贵的色彩搭配营造出奢华的氛围
3. 宾客可以坐在马毛皮椅上靠着壁炉小酌，壁炉两侧是手工雕刻的木制扩音器

1. The main attraction at Adventure Zone is its three heart-racing slides: Hyperglide Astra Slide; Double Drop Slide; and Tri-level Demon Drop Slide.
2. Age-appropriate play zones and games for toddlers and older children make play time at Adventure Zone a fun and thrilling experience for kids of all ages.
3. Adventure Zone also houses three festive party rooms. With themes such as Jingle Jungle, Pirate Ship and Circus Circus, they are the perfect party venues for the little ones.

1. 儿童冒险乐园主要拥有三个激动人心的滑梯：彩虹滑滑梯、双弯滑滑梯和魔鬼滑滑梯
2. 游戏区和游戏根据儿童的年龄设计，让各个阶段的儿童都能享有愉快的体验
3. 冒险乐园有三个派对房：丛林谷、海盗船和马戏团，十分适合儿童举办派对

1. The living room at Club Suite features reasonable and elegant layout of furnishing.
2. Club Premier Parkview Rooms offer great comfort in a spacious non-suite environment.
3. Style and spacious elegance define the Deluxe Rooms, located from levels 6 to 24. Guests can enjoy Century Park or city views.
4. Sleek and contemporary furnishings in Deluxe Premier provide optimum comfort for work, relaxation or socialising.
5. The unique blue sofa with silk pillows enhances the luxury and elegance of The Club Lounge.

1. 嘉里阁套房的客厅以理性而优雅的布局而著称
2. 嘉里阁超豪华园景客房为宾客提供宽敞舒适的非套房休憩空间
3. 豪华客房位于酒店6至24层，宽敞通透的格局洋溢着时尚优雅的气息。 放眼望去，世纪公园或繁华的都市风光一览无遗
4. 装潢时尚而现代的豪华客房为工作、休闲或社交最佳的舒适环境
5. 独特的蓝色沙发配有丝绸抱枕，凸显了嘉里阁行政酒廊的奢华和优雅

Index

索引

Accor

Sofitel Guangzhou Sunrich
Guangzhou, China
leesair@sina.com
CCD

Sofitel Macau at Ponte 16
Macau, China
H6480-SM3@sofitel.com
The Jerde Partnership, Richards Basmajian
Limited (Mansions in Sofitel Macau)

Sofitel Phnom Penh Phokeethra Resort & Spa
Phnom Penh, Cambodia
traineemgt1@sofitel-royal-angkor.com
Choochart Polakit

Starwood

The St. Regis Bangkok
Bangkok, Thailand
Jadechit.Khathathong@stregis.com
Brennan Beer Gorman Architects

The St. Regis Rome
Rome, Italy
Maddalena.Ciociola@starwoodhotels.com
Michael Stelea – HDC Interior Architecture +
Design, Architect Tomas Maier

The St. Regis Florence
Florence, Italy
Catarina.Pedras@starwoodhotels.com
Filippo Brunelleschi, Bottega Veneta

Sheraton Bangalore
Bangalore, India
info@dileonardo.com
Dileonardo

Sheraton Huizhou Beach Resort
Huizhou, China
bsdesign@163.com
Bangsheng Yang

The Peninsula

The Peninsula Hong Kong
Hong Kong, China
priscillachan@peninsula.com
Richmond International, Allison Henry
Designs, Rocco Design Limited

The Peninsula Tokyo
Tokyo, Japan
shunkikuchi@peninsula.com
Kazukiyo Sato, Yukio Hashimoto

The Peninsula Bangkok
Bangkok, Thailand
vpiyaphanee@peninsula.com
Glen Texeira Inc, Los Angeles; Denton
Corker Marshall Limited Hong Kong

InterContinental

InterContinental Puxi
Shanghai, China
honweng@ltwdesignworks.com
LTW Designworks Pte Ltd

InterContinental Regency Bahrain
Manama, Bahrain
nikki.b@dwp.com
Scott Whitaker, Kristina Zanic, Yutthana
Chanphong, Sue Henson (dwp)

Four Seasons

Four Seasons Los Angeles at Beverly Hills
Los Angeles, USA
sfadesign@sfadesign.com
SFA Design

Four Seasons Hotel Denver
Denver, USA
dana.berry@fourseasons.com
Bilkey Llinas Design, HKS, Inc., Carney Architects

Marriott

Ritz-Carlton Hong Kong
Hong Kong, China
honweng@ltwdesignworks.com
LTW Design works Pte Ltd

Ritz Carlton Dubai IFC
Dubai, United Arab Emirates
joshua.shi@ghcasia.com
Sandra M. Cortner (HBA)

JW Marriott Hotel Beijing
Beijing, China
elizabeth.xu@ghcasia.com
HBA

JW Marriott Marquis Miami
Miami, USA
WILee@RTKL.com
Wendy Mendes

Hilton

The Skirvin Hilton Oklahoma City
Oklahoma, USA
rfan@designdmu.com.cn
Turner Duncan, Kimberley Miller

Hilton Guangzhou Tianhe
Guangzhou, China
Jack.Tong@hilton.com
City Group

Hilton Chennai
Chennai, India
info@dileonardo.com
Andrew Chiu (DiLeonardo)

Waldorf-Astoria Shanghai on the Bund
Shanghai, China
joshua.shi@ghcasia.com
Ian Carr and Connie Puar (HBA)

Hyatt

Hyatt Regency Dusseldorf
Dusseldorf, Germany
g.dejong@fgstijl.nl
Sop architekten, FG stijl

Grand Hyatt Macau
Macau, China
macau.grand@hyatt.com
HBA

Hyatt Regency Jing Jin City Resort and Spa
Tianjin, China
jingjin.regency@hyatt.com
HEITZ PARSONS SADEK; Jean-Philippe Heitz

Park Hyatt Seoul
Seoul, South Korea
seoul.park@hyatt.com
Super Potato

Park Hyatt Beijing
Beijing, China
beijing.park@hyatt.com
Remedios Siembieda Inc, Super Potato & Bar Studio

Grand Hyatt Guangzhou
Guangzhou, China
guangzhou.grand@hyatt.com
Remedios Siembieda, Super Potato, Peter Remedios

Hyatt Regency Hong Kong, Tsim Sha Tsui
Hong Kong, China
hongkong.tsimshatsui@hyatt.com
EKIT II DESIGN CO., LTD.; Elvis Kwan

Park Hyatt Shanghai
Shanghai, China
shanghai.park@hyatt.com
Tony Chi & Associates

Hyatt Regency Hangzhou
Huangzhou, China
hangzhou.regency@hyatt.com
Florida Design Company, Heitz Parsons Sadek, Jean-Philippe Heitz

Hyatt Regency Hong Kong, Sha Tin
Hong Kong, China
hongkong.shatin@hyatt.com
Steve Leung

Mandarin Oriental

Mandarin Oriental, Tokyo
Tokyo, Japan
honweng@ltwdesignworks.com
LTW Designworks Pte Ltd

Mandarin Oriental, Singapore
Singapore
honweng@ltwdesignworks.com
LTW Designworks Pte Ltd

Mandarin Oriental, Boston
Boston, USA
fni@fnicholson.com
Frank Nicholson Incorporated, CBT Architects

The Landmark Mandarin Oriental, Hong Kong

Hong Kong, China

info@rsdesigners.com

Remedios Studio

Shangri-La

Shangri-La Hotel, Guangzhou

Guangzhou, China

slpg@shangri-la.com

HBA

Shangri-La Hotel, Wenzhou

Wenzhou, China

slgl@shangri-la.com

K.K.S Group

Shangri-La Hotel, Xian

Xian, China

slgl@shangri-la.com

Solari Design Limited

Shangri-La Hotel, Suzhou

Suzhou, China

slsz@shangri-la.com

L.R.F design

Shangri-La Hotel, Guilin

Guilin, China

slgl@shangri-la.com

Wong & Tung

Shangri-La Hotel Tokyo

Tokyo, Japan

elizabeth.xu@ghcasia.com

HBA

Shangri-La's Far Eastern Plaza Hotel, Tainan

Tainan, China

ivan.ng@edelman.com

Ed Ng, Terence Ngan (AB Concept)

Kerry Hotel Pudong, Shanghai

Shanghai, China

vivienne.tang@thekerryhotels.com

Kohn Pederson Fox Associates PC and Aedas
Ltd.; RA Design; Stickman Tribe Design

图书在版编目 (CIP) 数据

世界顶级酒店室内设计 /（荷）柯林·芬尼根编；
常文心译 . — 沈阳 : 辽宁科学技术出版社 , 2017.6
ISBN 978-7-5591-0130-3

Ⅰ . ①世… Ⅱ . ①柯… ②常… Ⅲ . ①饭店 – 室内
设计 – 图集 Ⅳ . ① TU238-64

中国版本图书馆 CIP 数据核字 (2017) 第 072578 号

出版发行：辽宁科学技术出版社
　　　　　（地址 : 沈阳市和平区十一纬路 25 号　邮编 : 110003）
印 刷 者：辽宁新华印务有限公司
经 销 者：各地新华书店
幅面尺寸：225mm × 305mm
印　　张：60
插　　页：4
字　　数：300 千字
出版时间：2017 年 6 月第 1 版
印刷时间：2017 年 6 月第 1 次印刷
责任编辑：李　红
封面设计：李　莹
版式设计：李　莹
责任校对：周　文

书　　号：ISBN 978-7-5591-0130-3
定　　价：398.00 元

编辑电话：024-23280367
邮购热线：024-23284502
E-mail: 1207014086@qq.com
http://www.lnkj.com.cn